Springer Undergraduate Mathematics Series

Springer

London
Berlin
Heidelberg
New York
Barcelona
Budapest
Hong Kong
Milan
Paris
Santa Clara
Singapore
Tokyo

D.A.R. Wallace

Groups, Rings and Fields

With 15 Figures

 Springer

D.A.R. Wallace, BSc, PhD, FRSE
Dept. of Mathematics, Strathclyde University, Livingstone Tower, 26 Richmond Street, Glasgow G1 1XH, UK

Cover illustration elements reproduced by kind permission of:
Aptech Systems, Inc., Publishers of the GAUSS Mathematical and Statistical System, 23804 S.E. Kent-Kangley Road, Maple Valley, WA 98038, USA. Tel: (206) 432 - 7855 Fax (206) 432 - 7832 email: info@aptech.com.URL: www.aptech.com
American Statistical Association: Chance Vol 8 No 1, 1995 article by KS and KW Heiner 'Tree Rings of the Northern Shawangunks' page 32 fig 2
Springer-Verlag: Mathematica in Education and Research Vol 4 Issue 3 1995 article by Roman E Maeder, Beatrice Amrhein and Oliver Gloor 'Illustrated Mathematics: Visualization of Mathematical Objects' page 9 fig 11, originally published as a CD ROM 'Illustrated Mathematics' by TELOS: ISBN 0-387-14222-3, german edition by Birkhauser: ISBN 3-7643-5100-4.
Mathematics in Education and Research Vol 4 Issue 3 1995 article by Richard J Gaylord and Kazume Nishidate 'Traffic Engineering with Cellular Automata' page 35 fig 2. Mathematica in Education and Research Vol 5 Issue 2 1996 article by Michael Trott 'The Implicitization of a Trefoil Knot' page 14.
Mathematica in Education and Research Vol 5 Issue 2 1996 article by Lee de Cola 'Coins, Trees, Bars and Bells: Simulation of the Binomial Process page 19 fig 3. Mathematica in Education and Research Vol 5 Issue 2 1996 article by Richard Gaylord and Kazume Nashidate 'Contagious Spreading' page 33 fig 1. Mathematica in Education and Research Vol 5 Issue 2 1996 article by Joe Buhler and Stan Wagon 'Secrets of the Madelung Constant' page 50 fig 1.

ISBN 3-540-76177-2 Springer-Verlag Berlin Heidelberg New York

British Library Cataloguing in Publication Data
Wallace, David A.R.
    Groups, rings and fields. – (Springer undergraduate mathematics series; 3423)
    1. Group theory   2. Rings (Algebra)   3. Algebraic fields
    I. Title
    512
ISBN 3540761772

Library of Congress Cataloging-in-Publication Data
Wallace, D.A.R. (David Alexander Ross)
    Groups, rings, and fields / D.A.R. Wallace.
       p.   cm. – (Springer undergraduate mathematics series)
    Includes index.
    ISBN 3-540-76177-2 (pbk. : alk. paper)
    1. Algebra, Abstract.   I. Title.   II. Series.
QA162.W36   1998   98-4961
512'.02–dc21   CIP

Typesetting by BC Typesetting, Bristol
Printed and bound at the Athenæum Press Ltd., Gateshead, Tyne & Wear
12/3830-54321 Printed on acid-free paper 10838358

Contents

Preface

This text is written for students who are meeting abstract algebra for the first time. Many of these students will take only one course of abstract algebra, whereas others may proceed to more advanced courses. The aim here is to provide an appropriate, interesting and entertaining text for those who require a rounded course as well as for those who wish to continue with further studies in algebra.

A fundamental difficulty for beginning students is often the axiomatic nature of abstract algebra and the exacting need to follow the axioms precisely. The present text is written so that an axiomatic treatment should seem to evolve as a natural development of intuitive ideas. To this end particular and extensive attention is paid to the integers, which are familiar objects of knowledge, and which, together with some simple properties of polynomials, are used to give motivation for the introduction of more abstract algebraic concepts. Historical allusions are made throughout the text in order to emphasise that abstract modern algebra has evolved from sometimes quite rudimentary ideas. Other remarks are intended to broaden, in a light-hearted manner, the student's general education.

Each section is provided with copious examples and ends with a set of exercises for which solutions are provided at the end of the book. In order to render the text suitable for self-study, any arguments in the text have been carefully crafted to facilitate the understanding, and to promote the enjoyment, of the reader. The text is self-contained and no prerequisites are absolutely necessary although a 'nodding acquaintance' with complex numbers and with matrices would, on occasion, be of advantage.

I must express my profound thanks to the members of the editorial team of Springer-Verlag and to my wife, Rachel Henderson, for her many encouragements and for her indefatigable efforts in turning a scribbled manuscript into an elegant word-processed document.

Finally, I am grateful to those correspondents who drew my attention to some textual peccadilloes, which have now been corrected.

D.A.R. Wallace
Department of Mathematics
University of Strathclyde
March 2001

1
Sets and Mappings

The notions of a 'set' and of a 'mapping' are fundamental in modern mathematics. In many mathematical contexts a perceptive choice of appropriate sets and mappings may lead to a better understanding of the underlying mathematical processes. We shall outline those aspects of sets and mappings which are relevant to present purposes and, for the delectation of the reader, conclude with a few logical paradoxes in regard to sets.

1.1 Union and Intersection

We are accustomed to speak of a 'collection' of books, or of an 'assembly' of people or of a 'list' of guests. In mathematics the word 'set' is used to denote the basic concept which is expressed by each of the collective nouns; in mathematics we say simply a 'set' of books, or a 'set' of people or a 'set' of guests.

Definition 1

A collection or assembly of objects is called a **set**. Each object is said to be an **element** of the set.

Thus in the phrase 'a set of books', each book involved in the set is an element of the set. Frequently the symbols constituting the elements of a (mathematical) set will be letters or numbers. If a set A consists of, say, the elements a, b, c then we write $A = \{a, b, c\}$, the use of curly brackets being customary. Notice that as we are concerned only with membership of the set we disregard any repetition of the symbols and have no preferred order for writing them down.

Example 1

Let A have 1, 2, 3, 4 as elements. Then $A = \{1, 2, 3, 4\} = \{1, 1, 2, 2, 3, 4\} = \{2, 1, 3, 4\} = \{1, 3, 4, 2\}$ etc.

Definition 2

Let A be a set. If A consists of a finite number of elements, say $a_1, a_2, \ldots, a_N$ (where the notation presumes the elements are distinct) then we write

$$A = \{a_1, a_2, \ldots, a_N\}$$

and A is said to be a **finite set** of **cardinality** N, written $N = |A|$. If A does not have a finite number of elements A is called an **infinite set** and is said to have **infinite cardinality**. The set consisting of no elements at all is called the **empty set** or the **null set** and is denoted by $\emptyset$ ($\emptyset$ is a letter of the Norwegian alphabet). Evidently $|\emptyset| = 0$.

Examples 2

1. $\emptyset$ = set of all unicorns.
2. $\emptyset$ = set of all persons living on the moon in 1996.
3. $|\{1, 2, 3, 4\}| = 4$.
4. $|\{a, b, c, d, e\}| = 5$.

A set is often given by some property which characterizes the elements of the set. Before giving a definition we offer a colourful example.

Example 3

Let A be the set of the colours of the rainbow. Whether or not a given colour is in A is determined by a property, namely that of being one of the colours of the rainbow. Of course in this particular instance we may write down the elements explicitly. Indeed

$$A = \{\text{red, orange, yellow, green, blue, indigo, violet}\}.$$

Definition 3

Let P be some property. The set of elements, each of which has the property P, is written as

$$A = \{x | x \text{ has the property } P\},$$

which we read as 'A is the set of all x such that x has the property P'.

Examples 4

1. Let A consist of the squares of strictly positive integers. Then

$$A = \{x | x = n^2, n = 1, 2, \ldots\}.$$

The set A may also be written as

$$A = \{1, 4, 9, \ldots\},$$

but notice that this notation is ambiguous since it would only be an inference that A consists of squares of integers.

2. The set $\{x | x < 0 \text{ and } x > 1\}$ is the empty set since there is no number which is simultaneously less than 0 and greater than 1.

3. The set $\{x | x^2 = 3, x \text{ is an integer}\}$ is empty since there is no integer with square equal to 3.

Definition 4

Let A be a set. Let x be an element. If x is **an element of** A we write

$$x \in A,$$

and if x is **not an element of** A we write

$$x \notin A.$$

Example 5

$$a \in \{a, b, c\}, \quad b \notin \{1, 2, 3\},$$

$$36 \in \{x | x = n^2, n = 0, 1, 2, \ldots\}, \quad 37 \notin \{x | x = n^2, n = 0, 1, 2, \ldots\}.$$

Definition 5

A set A is said to be a **subset** of a set B if every element of A is also an element of B. We write

$$A \subseteq B$$

to indicate that A is a **subset** of B and read the notation as 'A is contained in or equal to B'. The empty set is deemed to be a subset of every set. If A is a subset of B but $A \neq B$ then A is said to be a **proper** subset of B.

If A is not a subset of B then there exists at least one element of A which is not an element of B and we write

$$A \nsubseteq B.$$

If $A \subseteq B$ and $B \subseteq A$ then the sets A and B are **equal** and we write

$$A = B.$$

If A and B are **not equal** we write

$$A \neq B.$$

If $A \subseteq B$ and $A \neq B$ we write $A \subset B$.

Given two sets A and B, to prove that $A \subseteq B$ we have to show that $x \in A$ implies $x \in B$. To prove that $A \nsubseteq B$, we have to exhibit an $x \in A$ such that $x \notin B$.

Two sets A and B are equal if and only if they are elementwise indistinguishable.

Examples 6

1. Let $A = \{a, c\}, B = \{a, b, c, d\}, C = \{a, b, c\}$. Then $A \subseteq B, C \subseteq B$.
2. Let $A = \{x | x^2 = 1\}, B = \{-1, 1\}$. Although A and B are described in different ways they have the same elements and so $A = B$.

Definition 6

Applications of the theory of sets usually take place within some fixed set (which naturally varies according to the circumstances). This fixed set for the particular application is called the **universal set** and is often denoted by U. The subset of U consisting of the elements of U having the **property** P is denoted by

$$\{x \in U | x \text{ has the property } P\},$$

or simply as

$$\{x | x \text{ has the property } P\}$$

if the universal set is obvious. Frequently explicit reference to a universal is omitted.

We now introduce the operations of union and intersection on sets.

Definition 7

Let $A, B, C, \ldots$ be sets. The set consisting of those elements each of which is in at least one of the sets $A, B, C, \ldots$ is called the **union** of $A, B, C, \ldots$ and is written

$$A \cup B \cup C \cup \ldots$$

The set consisting of those elements each of which is in all of the sets $A, B, C, \ldots$ is called the **intersection** of $A, B, C, \ldots$ and is written as

$$A \cap B \cap C \cap \ldots$$

The sets $A, B, C, \ldots$ are said to be **disjoint** if the intersection of any two distinct sets is empty, that is, $A \cap B = A \cap C = B \cap C = \ldots = \varnothing$.

The union $A \cup B \cup C \cup \ldots$ is said to be a **disjoint union** if the sets $A, B, C, \ldots$ are disjoint.

Example 7

Let $U = \{a, b, c, d, e, f, g, h, i\}$ be the universal set. Let $A = \{a, b, c\}$, $B = \{d, f, i\}$, $C = \{a, b, c, d, f, g\}$, $D = \{a, e, g, h, i\}$, $E = \{e, g, h\}$.

Then

$$A \cup B = \{a, b, c, d, f, i\}, \quad A \cap B = \varnothing,$$

$$C \cup D = \{a, b, c, d, e, f, g, h, i\} = U, \quad C \cap D = \{a, g\},$$

$$A \cup B \cup C = \{a, b, c, d, f, g, i\}, \quad A \cap B \cap C = \varnothing,$$

$$B \cup C \cup D = \{a, b, c, d, e, f, g, h, i\} = U, \quad B \cap C \cap D = \varnothing,$$

$$U = A \cup B \cup E,$$

where this union is disjoint as $A \cap B = A \cap E = B \cap E = \varnothing$.

It is immediate that for sets A and B we have $A \cap A = A$, $A \cup A = A$ and $A \cap B = B \cap A \subseteq A \subseteq A \cup B = B \cup A$.

We prove a useful lemma.

Lemma 1

Let A, B, X be sets such that $A \subseteq B$. Then the following hold.

1. $A \cap X \subseteq B \cap X$.
2. $A \cup X \subseteq B \cup X$.

Proof

1. We have to prove that every element of $A \cap X$ is also in $B \cap X$.
   Let $c \in A \cap X$. Then $c \in A$ and so, as $A \subseteq B$, $c \in B$. But $c \in X$ and so $c \in B \cap X$. Hence $A \cap X \subseteq A \cap X$.
2. Let $c \in A \cup X$. Then either $c \in A$ or $c \in X$. If $c \in A$ then $c \in B$ and so, in either case, $c \in B \cup X$. Thus $A \cup X \subseteq B \cup X$. $\qquad\square$

Definition 8

Let A be a subset of the universal set U. The subset of U consisting of all elements of U not in A is called the **complement** (note spelling!) of A. The complement of A is denoted by $U \setminus A$ or by A' if the universal set is clear.

$$U \setminus A = A' = \{x \in U | x \notin A\}$$

(Another notation for the complement, but not used in this text, is $\mathbb{C}A$.)

$$U = (U \setminus A) \cup A, \; (U \setminus A) \cap A = \varnothing,$$

and so

$$A = U \text{ if and only if } A' = \varnothing.$$

Example 8

Let $U = \{1, 2, 3, 4, 5\}$, $A = \{2, 3\}$. Then $U \setminus A = \{1, 4, 5\}$.

The notion of complement extends to that of relative complement.

Definition 9

Let A and B be sets. The **relative complement** of A in B, denoted by $B \setminus A$, is the set of elements of B which are not in A.

Example 9

Let $A = \{p, q, r\}$, $B = \{q, r, s, t\}$. Then $B \setminus A = \{s, t\}$, $A \setminus B = \{p\}$.

We conclude this section with a fairly obvious but useful result.

Theorem 1

Let A and B be sets. Then the following statements are equivalent.

1. $A \subseteq B$.
2. $A \cap B = A$.
3. $A \cup B = B$.

Proof

We prove the equivalence of 1 and 2, leaving the equivalence of 1 and 3 as an exercise.

Suppose $A \subseteq B$. Then, by Lemma 1,

$$A = A \cap A \subseteq A \cap B \subseteq A, \text{ and thus}$$

$$A = A \cap B.$$

Conversely if $A = A \cap B$ then certainly $A \subseteq B$. $\square$

We conclude this section by introducing the 'Cartesian product' of sets. The adjective 'Cartesian' derives from the name of R. du P. Descartes (1596–1650), mathematician and philosopher whose philosophical outlook is enshrined in his famous aphorism "je pense, donc je suis", in English "I think, therefore I am". (Quotation from Discours de la Methode [Leyden, 1637, 4th part]).

Definition 10

Let A and B be two sets. The **Cartesian product** $A \times B$ of A and B is defined to be the set of all ordered pairs of the form $(a, b), a \in A, b \in B$;

$$A \times B = \{(a, b) | a \in A, b \in B\}.$$

The **Cartesian product** $A_1 \times A_2 \times \ldots \times A_n$ of the sets $A_1, A_2, \ldots, A_n$ (in this order) is defined to be the set of all ordered n-tuples of the form

$$(a_1, a_2, \ldots, a_n), a_1 \in A_1, a_2 \in A_2, \ldots, a_n \in A_n,$$

$$A_1 \times A_2 \times \ldots \times A_n = \{(a_1, a_2, \ldots, a_n) | a_i \in A_i, i = 1, 2, \ldots, n\}.$$

Note that if the sets A and B are distinct, then $A \times B \neq B \times A$.

Example 10

Let A and B be finite sets of m and n elements respectively. Then $A \times B$ is a set of mn elements since in the ordered pair (a, b) there are m possibilities for a, and n possibilities for b.

Exercises 1.1

1. Let the universal set U be the set $\{u, v, w, x, y, z\}$. Let $A = \{u, v, w\}$, $B = \{w, x, y\}$, $C = \{x, y, z\}$. Write down the subsets: $A \cup B$, $A \cap B$, $A \cap C$, $A \cup C$, $A \setminus B$, $B \setminus A$, $A \setminus C$, $C \setminus A$, A', B', C', $A \setminus (B \cup C)$, $B \setminus (A \cap C)$, $(A \cap B) \cup (B \setminus A)$.

2. Let the universal set U be the set $\{a, b, c, 1, 2, 3\}$. Let $X = \{a, b, 1\}$, $Y = \{b, 2, 3\}$, $Z = \{c, 2, 3\}$. Write down the subsets: $X \cup Y$, $X \cap Y$, $X \cup Z$, $X \cap Z$, $X \setminus Y$, $Y \setminus X$, $X \setminus Z$, $Z \setminus X$, X', Y', Z', $(X \cap Y)'$, $X' \cup Y'$, $(X \cup Z)'$, $X' \cap Z'$.

3. Let the universal set U be the set $\{1, 2, 3, 4, 5, 6\}$. Let $A = \{1, 2, 4, 5\}$. Let X be a subset of U such that $A \cup X = \{1, 2, 4, 5, 6\}$ and $A \cap X = \{2, 4\}$. Prove that there is only one possibility for X, namely $\{2, 4, 6\}$.

4. Let A and B be sets. Prove that $A \subseteq B$ if and only if $B = A \cup B$ (Theorem 1).

5. Let A be the set $\{3, 5, 8, \dots\}$. Is A the set $\{3, 5, 8, 13, \dots\}$ or the set $\{3, 5, 8, 12, \dots\}$ or neither?

6. Let A_i be a finite set of k_i elements, $i = 1, 2, \dots, n$. Prove that
$$A_1 \times A_2 \times \dots \times A_n \text{ is a finite set of } k_1 k_2 \dots k_n \text{ elements.}$$

1.2 Venn Diagrams

In this section we shall describe a technique which, in certain circumstances, enables us to visualize sets and their unions and intersections etc. by drawing pictures on paper. The resulting pictures are known as **Venn diagrams**, after J. Venn (1834–1923).

We proceed as follows. For the universal set U we draw a rectangle and for arbitrary subsets A, B of U we draw shapes as below:

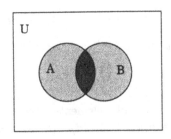

Figure 1.1.

The shaded area represents $A \cup B$ and the dark-shaded area $A \cap B$. Conveniently we may assign numbers to the regions:

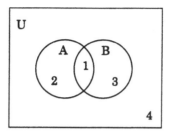

Figure 1.2.

Thus

region 1 represents $A \cap B$,

region 2 represents $A \setminus (A \cap B)$,

region 3 represents $B \setminus (A \cap B)$,

region 4 represents $U \setminus (A \cup B)$.

The Venn diagram makes clear that A is the disjoint union of the two subsets $A \setminus B$ and $A \cap B$, and $A \cup B$ is the disjoint union of the three subsets $A \setminus B$, $B \setminus A$ and $A \cap B$.

Wise use of Venn diagrams may lead to general results for which a rigorous proof may later be obtained. We give two examples of this procedure in the following examples.

Example 11

Using the above Venn diagram we may observe the following:

A	is represented by regions		1, 2
A'	... ...	...	3, 4
B	... ...	...	1, 3
B'	... ...	...	2, 4
$(A \cap B)'$	... ...	...	2, 3, 4
$A' \cup B'$	... ...	...	2, 3, 4

Thus, from our Venn diagram representation, we conclude that

$$(A \cap B)' = A' \cup B'.$$

Similarly we conclude that

$$(A \cup B)' = A' \cap B'.$$

A rigorous proof will be given in Theorem 3.

If now C is a third subset of U then the Venn diagram becomes:

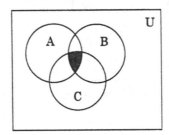

Figure 1.3.

The shaded region represents $A \cap B \cap C$.

For four or more subsets of U the method has fundamental limitations (as those attempting to draw satisfactory diagrams of four subsets will discover) and should be avoided.

Example 12

Let A, B, C be subsets of the universal set U and suppose $A \subseteq C$, $B \subseteq C$. The appropriate Venn diagram is:

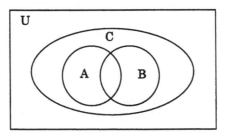

Figure 1.4.

Example 13

Let us draw a Venn diagram to illustrate the result that if $X \subseteq Y$ then $A \cap X \subseteq A \cap Y$.

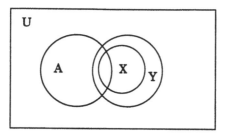

Figure 1.5.

Consider a Venn diagram with three subsets A, B, C of the universal set U and with regions assigned numbers as follows:

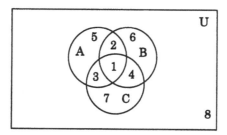

Figure 1.6.

region 1 represents $A \cap B \cap C$

region 2 represents $(A \cap B) \setminus (A \cap B \cap C)$

region 3 represents $(A \cap C) \setminus (A \cap B \cap C)$

region 4 represents $(B \cap C) \setminus (A \cap B \cap C)$

$\cdots \qquad \cdots \qquad \cdots \qquad \cdots \qquad \cdots$

Example 14

In the above diagram, suppose we are asked to determine the regions corresponding to the following subsets:

$$B \cup C, A \cap (B \cup C), A \cap B, A \cap C, (A \cap B) \cup (A \cap C).$$

Then

$B \cup C$	is represented by regions	1, 2, 3, 4, 6, 7
$A \cap (B \cup C)$	$\cdots \quad \cdots \quad \cdots$	1, 2, 3
$A \cap B$	$\cdots \quad \cdots \quad \cdots$	1, 2
$A \cap C$	$\cdots \quad \cdots \quad \cdots$	1, 3
$(A \cap B) \cup (A \cap C)$	$\cdots \quad \cdots \quad \cdots$	1, 2, 3

We notice that, from our Venn diagram representation,

$$A \cap (B \cup C) = (A \cap B) \cup (A \cap C).$$

Similarly we may conclude that

$$A \cup (B \cap C) = (A \cup B) \cap (A \cup C).$$

A rigorous proof is given in Theorem 2.

Theorem 2

Let A, B, C be sets. Then

1. $A \cap (B \cup C) = (A \cap B) \cup (A \cap C)$
2. $A \cup (B \cap C) = (A \cup B) \cap (A \cup C)$

(These results are called the **Distributive Laws** for sets.)

Proof

We prove 1, the proof of 2 is similar.

Since $B \subseteq B \cup C$ and $C \subseteq B \cup C$, we have $A \cap B \subseteq A \cap (B \cup C)$ and $A \cap C \subseteq A \cap (B \cup C)$ from which

$$(A \cap B) \cup (A \cap C) \subseteq A \cap (B \cup C).$$

We now have to prove that $A \cap (B \cup C) \subseteq (A \cap B) \cup (A \cap C)$. For this part of the proof we consider elements and not simply subsets. Therefore let $x \in A \cap (B \cup C)$. Then $x \in A$ and $x \in B \cup C$ and so x is in either B or C. If we suppose $x \in B$ then $x \in B$ and $x \in A$ from which we have $x \in A \cap B$ and so $x \in (A \cap B) \cup (A \cap C)$. The alternative supposition that $x \in C$ leads to the same conclusion. Consequently we have

$$A \cap (B \cup C) \subseteq (A \cap B) \cup (A \cap C)$$

This completes the proof.   □

The next result (see Example 11) gives the so-called De Morgan Laws, after A. De Morgan (1806–71).

Theorem 3 De Morgan Laws

Let A, B be sets. Then the following statements hold.

1. $(A \cap B)' = A' \cup B'$.
2. $(A \cup B)' = A' \cap B'$.

Proof

We prove 1, the proof of 2 is similar.

Let $x \in (A \cap B)'$. Then $x \notin (A \cap B)$. This means that $x \notin A$ or $x \notin B$. If $x \notin A$ then $x \in A'$ and so $x \in A' \cup B'$. If $x \notin B$ then $x \in B'$ and so $x \in A' \cup B'$. Thus $(A \cap B)' \subseteq A' \cup B'$.

Let $x \in A' \cup B'$. Then $x \in A'$ or $x \in B'$. If $x \in A'$ then $x \notin A$ and so certainly $x \notin A \cap B$ and similarly if $x \in B'$ then $x \notin A \cap B$. Thus $x \notin A \cap B$ and so $x \in (A \cap B)'$. Thus $A' \cup B' \subseteq (A \cap B)'$. Hence we obtain the desired result. $\square$

Many of the sets with which we shall be concerned will be infinite, but for finite sets we now introduce some counting techniques.

If A and B are both finite sets then certainly $A \cup B$ is finite since there cannot be more elements than in A and B considered separately. We shall derive a formula for the number of distinct elements in $A \cup B$. We recall that $|A|$ is the number of elements in A (Definition 2).

Theorem 4

Let A and B be finite sets. Then $A \cup B$ is a finite set and

$$|A \cup B| = |A| + |B| - |A \cap B|.$$

Proof

Let us examine the Venn diagram for $A \cup B$. As we remarked

$$A = (A \setminus B) \cup (A \cap B),$$
$$B = (B \setminus A) \cup (A \cap B),$$
$$A \cup B = (A \setminus B) \cup (B \setminus A) \cup (A \cap B)$$

are disjoint unions of subsets of A or B. These subsets are necessarily finite and so, by obvious (and legitimate) counting, we have

$$|A| = |A \setminus B| + |A \cap B|,$$
$$|B| = |B \setminus A| + |A \cap B|$$

and also

$$|A \cup B| = |A \setminus B| + |B \setminus A| + |A \cap B|$$
$$= [|A| - |A \cap B|] + [|B| - |A \cap B|] + |A \cap B|$$
$$= |A| + |B| - |A \cap B|. \qquad \square$$

The following two examples illustrate the use of this formula.

Example 15

Let $A = \{a, b, c, d\}$, $B = \{a, c, e, f, g\}$.
Then $A \cup B = \{a, b, c, d, e, f, g\}$, $A \cap B = \{a, c\}$. As we expect

$$|A \cup B| = 7 = 4 + 5 - 2 = |A| + |B| - |A \cap B|.$$

Example 16

We are told that in a party of 95 English-speaking schoolchildren there are 20 who speak only English, 60 who can speak French and 24 who can speak German. We require to determine how many speak both French and German.

The 'universal set' consists of 95 schoolchildren. We let F and G be the subsets consisting of those schoolchildren who can speak French and German respectively. Then $|F| = 60$, $|G| = 24$. $F \cup G$ is the subset of those schoolchildren who speak either French or German. Since 20 speak only English we have

$$|F \cup G| = 95 - 20 = 75.$$

$$\text{Then } |F \cup G| = |F| + |G| - |F \cap G|$$

$$\text{and so } 75 = 60 + 24 - |F \cap G|.$$

$$\text{Thus } |F \cap G| = 9.$$

This says that precisely 9 schoolchildren can speak both French and German.

We may extend Theorem 4 to consider three or more sets, but we confine our extension to the case of three sets.

Theorem 5

Let A, B and C be finite sets. Then $A \cup B \cup C$ is a finite set and

$$|A \cup B \cup C| = |A| + |B| + |C| - |A \cap B| - |A \cap C| - |B \cap C| + |A \cap B \cap C|.$$

Proof

We apply Theorem 4. We note that

$$(A \cap B) \cap (A \cap C) = A \cap B \cap C.$$

Then, on letting $P = B \cup C$, we have

$$|A \cup B \cup C| = |A \cup P|$$
$$= |A| + |P| - |A \cap P|$$
$$= |A| + |B \cup C| - |A \cap P|$$
$$= |A| + |B| + |C| - |B \cap C| - |A \cap (B \cup C)|$$
$$= |A| + |B| + |C| - |B \cap C| - |(A \cap B) \cup (A \cap C)| \text{ (by Theorem 2)}$$
$$= |A| + |B| + |C| - |B \cap C| - [|A \cap B| + |A \cap C|$$
$$- |(A \cap B) \cap (A \cap C)|]$$
$$= |A| + |B| + |C| - |A \cap B| - |A \cap C| - |B \cap C| + |A \cap B \cap C|.$$

$\square$

Applications of this result are often made to problems of which the following is typical.

Example 17

In a gathering of 136 fairly athletic students, 92 engage in gymnastics, 68 are swimmers and 78 are tennis players. Of the 136 students, 41 do gymnastics and swim, 43 do gymnastics and play tennis and 24 swim and play tennis. If 4 students participate neither in gymnastics, swimming nor tennis, how many indulge in all three of these sporting activities? How many play tennis but do not have other sporting activities?

We have to reduce the situation of the problem to manageable mathematics.

Let U be the set of 136 students,

let G be the set of gymnasts,

let S be the set of swimmers, and

let T be the set of tennis players.

We are given the information that

$$|G| = 92, |S| = 68, |T| = 78, |G \cap S| = 41, |G \cap T| = 43, |S \cap T| = 24.$$

We are also given that $|U \setminus (G \cup S \cup T)| = 4$ from which we deduce that

$$|G \cup S \cup T| = |U| - |U \setminus (G \cup S \cup T)|$$
$$= 136 - 4 = 132$$

Now applying Theorem 5 we have

$$132 = |G \cup S \cup T| = |G| + |S| + |T| - |G \cap S| - |G \cap T| + |G \cap S \cap T|$$
$$= 92 + 68 + 78 - 41 - 43 - 24 + |G \cap S \cap T|.$$

Hence $|G \cap S \cap T| = 2$, and so 2 students engage in all three activities.

The number of students who play tennis but have no other sporting activities is given by $|T \setminus (G \cup S)|$. The easiest way to determine this number is by means of a Venn diagram in which $|G \cap S \cap T| = 2$, a is the number of students who are gymnasts and play tennis but do not swim, and b is the number of students who swim and play tennis but are not gymnasts.

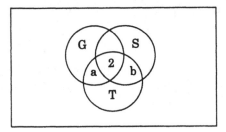

Figure 1.7.

We have

$$a = |(G \cap T) \setminus S| = |(G \cap T) \setminus (G \cap S \cap T)|$$
$$= |G \cap T| - |G \cap S \cap T|$$
$$= 43 - 2$$
$$= 41.$$

Similarly

$$b = |(S \cap T) \setminus G|$$
$$= 24 - 2$$
$$= 22.$$

Then, from the diagram, the number of students who play tennis but have no other sports is

$$|T| - a - b - 2 = 78 - 41 - 22 - 2$$
$$= 13.$$

Exercises 1.2

1. Let A and B be sets. Establish the De Morgan Law
$$(A \cup B)' = A' \cap B'$$
   (i) by a Venn diagram,
   (ii) by a set-theoretic proof.

2. Let A, B and C be sets. Establish the Distributive Law
$$A \cup (B \cap C) = (A \cup B) \cap (A \cup C)$$
   (i) by a Venn diagram,
   (ii) by a set-theoretic proof.

3. Let A, B and C be sets. Prove, by constructing an appropriate example (often called a counter-example) that the result
$$A \cup (B \cap C) = (A \cup B) \cap C$$
   does not necessarily hold. If $A \subseteq C$ does the result hold? Is the condition $A \subseteq C$ necessary and sufficient for the result to hold?

4. In a certain village of 1000 well-educated people, 250 read the *Economist*, 411 read the *Bangkok Post* and 315 read the *Straits Times*. Of these people, 72 read the *Economist* and the *Bangkok Post*, 51 read the *Economist* and the *Straits Times* and 31 read the *Bangkok Post* and the *Straits Times*. If only one person reads all three of the publications, how many do not read any of the publications?

5. A party of 200 schoolchildren is investigated with regard to likes and dislikes of three items, namely, ice cream, sweets and fizzy lemonade. It is found that of those children who like ice cream only 7 dislike sweets. 110 children like sweets and 149 like fizzy lemonade. 80 children like both sweets and fizzy lemonade and 36 like both ice cream and lemonade. If 31 children consume all three items avidly, how many children like none of these items?

1.3 Mappings

In our mathematics we have probably evaluated expressions such as x^2, $\sin x$, and $\sqrt{x^2 - 4}$ for given 'values of x'. We may have described these expressions loosely as being functions of x. While this description is somewhat vague it does encapsulate a concept which we would wish to make rather more precise. For a given value of x we expect unique evaluations of the expressions; thus

$(-2)^2 = 4, \sin \pi/4 = 1/\sqrt{2}$, but it is not permissible to put $x = 1$ in $\sqrt{x^2 - 4}$ if we wish to obtain a real square root. This suggests that we have to specify how the 'values of x' may be chosen. We begin to appreciate that, in each case, we need to specify a set of elements and to have the function defined exactly on this set. We make the following definition.

Definition 11

Let A and B be sets. Let there be a rule or prescription, denoted by f, by which to each element a of A there is assigned a unique element, denoted by $f(b)$, of B. Then the rule is said to be a **mapping** or **map** or **function** from A to B. We write

$$a \to f(a) \quad (a \in A)$$

and, to indicate that f is a mapping on sets, we use both of the following notations:

$$f : A \longrightarrow B \text{ and } A \xrightarrow{f} B.$$

A is called the **domain** of f and B is called the **codomain** of f.

[The designations 'map' and 'function' are perhaps more common in topology and analysis respectively. We shall use the term 'mapping'.]

For $a \in A$, $f(a)$ is called the **image** of a and, likewise, the subset of B given as

$$\{f(a)|a \in A\}$$

is called the **image** of A or, sometimes, the **range** of f. We also denote this subset by $f(A)$. Finally we note that $f(\varnothing) = \varnothing$.

Example 18

Let $A = \{a, b, c\}$ and $B = \{0, 1\}$. We define a mapping f where $f : A \to B$ by letting

$$f(a) = 0, f(b) = 1, f(c) = 1.$$

Then f has domain A, codomain B and range B. We define a second mapping g where $g : A \to B$ by letting

$$g(a) = 1, g(b) = 1, g(c) = 1.$$

Then g has domain A, codomain B and range $\{1\}$. We note that f and g are unequal since, for example,

$$f(a) = 0 \neq 1 = g(a).$$

The condition under which two mappings are equal should be fairly obvious but, for the sake of completeness, we state the condition formally.

Remark

Two mappings f and g are **equal** if and only if f and g have the same domain A and $f(a) = g(a)$ for all $a \in A$.

Example 19 (Continued from Example 3)

Let A be the set of the colours of the rainbow. Let $B = \{1, 2, 3, 4, 5, 6, 7\}$. We define a mapping f where $f : A \to B$ by the rule that the image of a given colour of the rainbow is the number of letters in that colour. We need not go further to specify the mapping although, in fact, we have

$$f(\text{red}) = 3, f(\text{orange}) = 6, f(\text{yellow}) = 6, f(\text{green}) = 5,$$

$$f(\text{blue}) = 4, f(\text{indigo}) = 6, f(\text{violet}) = 6.$$

We now introduce certain important sets of which we shall have more to say in the next chapter.

Notation

The set of **natural numbers**, which is also the set of **strictly positive integers**, is denoted by $\mathbb{N}$ ($\mathbb{N}$ for 'natural'),

$$\mathbb{N} = \{1, 2, 3, \ldots\}.$$

The set of **integers** is denoted by $\mathbb{Z}$ ($\mathbb{Z}$ for 'Zahl', the German word for 'number'),

$$\mathbb{Z} = \{\ldots, -2, -1, 0, 1, 2, \ldots\}.$$

The set of **rational numbers**, which is also the set of quotients of integers, is denoted by $\mathbb{Q}$ ($\mathbb{Q}$ for 'quotient'),

$$\mathbb{Q} = \{m/n \mid m, n \in \mathbb{Z}, n \neq 0\}.$$

Thus $\mathbb{Q}$ consists of all fractions such as $\frac{1}{2}, \frac{2}{3}, \frac{-15}{8}$.

The set of **real numbers** is denoted by $\mathbb{R}$ and the set of **complex numbers** is denoted by $\mathbb{C}$,

$$\mathbb{C} = \{a + ib : a, b \in \mathbb{R}\}.$$

Evidently

$$\mathbb{N} \subseteq \mathbb{Z} \subseteq \mathbb{Q} \subseteq \mathbb{R} \subseteq \mathbb{C}.$$

A real number which is not rational is said to be **irrational**, $\mathbb{R} \setminus \mathbb{Q}$ is the set of irrational numbers.

We give three examples of mappings involving these sets before giving some useful terminology for describing mappings.

Examples 20

1. The mapping $f : \mathbb{N} \to \{-1, 1\}$ given by
$$f(n) = (-1)^n \quad (n \in \mathbb{N})$$
    has domain $\mathbb{N}$, codomain $\{-1, 1\}$ and range $\{-1, 1\}$.

2. The mapping $f : R \to R$ given by
$$f(x) = x^2 + 1 \quad (x \in \mathbb{R})$$
    has domain $\mathbb{R}$, codomain $\mathbb{R}$ and range $\{y | y \geq 1\}$.

3. The mapping $f : \mathbb{N} \to \mathbb{Q}$ given by
$$f(n) = \tfrac{1}{2}n \quad (n \in \mathbb{N})$$
    has domain $\mathbb{N}$, codomain $\mathbb{Q}$ and range $\{\tfrac{1}{2}, 1, \tfrac{3}{2}, 2, \ldots\} = \mathbb{N} \cup \{\tfrac{n}{2} | n \in \mathbb{N}\}$.

Definition 12

Let A and B be sets. Let $f : A \to B$ be a mapping.

(i)   The mapping f is said to be **injective,** or **one–one**, if whenever $f(a_1) = f(a_2)$ for $a_1, a_2 \in A$ then necessarily $a_1 = a_2$.

(ii) The mapping f is said to be **surjective** or **onto** if for each $b \in B$ there exists $a \in A$ such that $f(a) = b$.

(iii) The mapping f is said to be **bijective** or **one–one and onto** if f is both injective and surjective.

Examples 21

1. Let $A = \{a, b, c\}$ and $B = \{p, q\}$. We define $f : A \to B$ by
$$f(a) = p, \ f(b) = p, \ f(c) = p.$$
    Then f is not injective since $f(a) = f(b)$ but $a \neq b$. We also have that f is not surjective since $f(A) = \{p\} \neq B$.

2. Let $A = \{a, b\}$ and $B = \{p, q, r\}$. We define $f : A \to B$ by
$$f(a) = p, \ f(b) = q.$$
    Then f is injective since $f(a) \neq f(b)$ but is not surjective since $f(A) = \{p, q\} \neq B$.

3. Let $A = \{a, b, c\}$ and $B = \{p, q\}$. We define
$$f : A \to B \text{ by } f(a) = p, \ f(b) = p, \ f(c) = q.$$
    Then f is not injective since $f(a) = f(b)$ but is surjective since $f(A) = \{p, q\} = B$.

[The reader may begin to suspect (correctly) that a bijective mapping exists between two finite sets if and only if they have the same number of elements. In particular the reader may rightly conclude that a bijective mapping cannot exist between a finite set and a proper subset; for infinite sets the situation is less restrictive, as the next example shows.]

4. Let 2N denote the set of strictly positive even integers,

$$2N = \{2n | n \in N\}.$$

Now 2N is a proper subset of N but, nevertheless, the mapping $f : N \to 2N$ given by

$$f(n) = 2n \quad (n \in N)$$

is bijective.

5. A mapping f is defined on N by

$$f(n) = \begin{cases} n & (n \text{ odd}) \\ \frac{1}{2}n & (n \text{ even}). \end{cases}$$

Suppose we are asked to describe f. We note first that f is defined so that $f : N \to N$. Then f is not injective since, for example, $f(3) = 3 = f(6)$. But f is surjective since if $k \in N$ then $f(2k) = k$.

Frequently we have to consider the effect of applying mappings successively. More precisely we have the following definition.

Definition 13

Let A, B and C be sets and let $f : A \to B$ and $g : B \to C$ be given mappings. We have

$$A \xrightarrow{f} B \xrightarrow{g} C$$

where the range of f is a subset of the domain of g. Then for each $a \in A$ we have a mapping, denoted by $g \circ f$, called the **circle-composition** of the mappings g and f and given by

$$(g \circ f)(a) = g(f(a))$$

in which we first apply the mapping f to $a \in A$ and then apply the mapping g to $f(a) \in B$. If no ambiguity will arise we often write gf for $g \circ f$.

Theorem 6

Let A, B and C be sets and let $f : A \to B$ and $g : B \to C$ be mappings.

1. If f and g are surjective then $g \circ f$ is surjective.
2. If f and g are injective then $g \circ f$ is injective.
3. If f and g are bijective then $g \circ f$ is bijective.

Proof

We prove 1 and leave 2 to be proved in subsequent Exercises. 1 and 2 imply 3. Let $c \in C$. Then since g is surjective we have $b \in B$ such that $g(b) = c$. Since f is surjective we have $a \in A$ such that $f(a) = b$. Then

$$(g \circ f)(a) = g(f(a)) = g(b) = c$$

and so $g \circ f$ is surjective.   □

Examples 22

1. Let $A = \{a, b, c, d\}$, $B = \{1, 2, 3\}$, $C = \{p, q, r, s\}$. Let mappings f, g be defined by $f(a) = 1$, $f(b) = 1$, $f(c) = 2$, $f(d) = 3$, $g(1) = p$, $g(2) = r$, $g(3) = r$. Then we have

$$A \xrightarrow{f} B \xrightarrow{g} C$$

with, for example,

$$(g \circ f)(b) = g(f(b)) = g(1) = p,$$

$$(g \circ f)(d) = g(f(d)) = g(3) = r.$$

2. Let mappings f and g be defined by

$$f(x) = \sin x \quad (x \in \mathbb{R}), \qquad g(x) = x^2 \quad (x \in \mathbb{R}).$$

Then we may form $g \circ f$ and also $f \circ g$:

$$\mathbb{R} \xrightarrow{f} \mathbb{R} \xrightarrow{g} \mathbb{R},$$

$$\mathbb{R} \xrightarrow{g} \mathbb{R} \xrightarrow{f} \mathbb{R},$$

Now

$$(g \circ f)(x) = g(f(x)) = g(\sin x) = (\sin x)^2 = \sin^2 x$$

but

$$(f \circ g)(x) = f(g(x)) = f(x^2) = \sin x^2.$$

$g \circ f$ and $f \circ g$ have the same domain but are different mappings since it is not true, contrary to an occasional misguided belief, that $\sin^2 x = \sin x^2$ for all $x \in \mathbb{R}$.

Consider the case of four sets A, B, C, D and three mappings f, g, h where $f : A \to B$, $g : B \to C$ and $h : C \to D$.

We may construct the mappings $g \circ f$ and $h \circ g$ and then the further compositions $h \circ (g \circ f)$ and $(h \circ g) \circ f$, giving the pictures below.

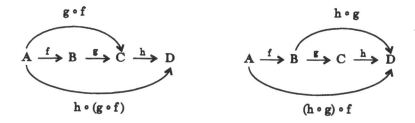

Figure 1.8.

Then $h \circ (g \circ f)$ and $(h \circ g) \circ f$ are both mappings from A to D but they have been constructed differently. Could they be equal?

We consider a specific example before proving the general result.

Example 23

Let f, g, h be the mappings given by

$$f(x) = x^2 + 1 \quad (x \in \mathbb{Z}),$$

$$g(n) = \frac{2}{3}n \quad (n \in \mathbb{N}),$$

$$h(t) = \sqrt{t^2 + 1} \quad (t \in \mathbb{Q}),$$

and so,

$$\mathbb{Z} \xrightarrow{\ f\ } \mathbb{N} \xrightarrow{\ g\ } \mathbb{Q} \xrightarrow{\ h\ } \mathbb{R}$$

But

$$(g \circ f)(x) = g(f(x)) = g(x^2 + 1) = \frac{2}{3}(x^2 + 1) \quad (x \in \mathbb{Z}),$$

$$[h \circ (g \circ f)](x) = h\left(\frac{2}{3}(x^2 + 1)\right) = \sqrt{\left(\frac{2}{3}(x^2 + 1)\right)^2 + 1},$$

$$(h \circ g)(n) = h(g(n)) = h\left(\frac{2}{3}n\right) = \sqrt{\left(\frac{2}{3}n\right)^2 + 1} \quad (n \in \mathbb{N}),$$

$$[(h \circ g) \circ f](x) = (h \circ g)(f(x)) = (h \circ g)(x^2 + 1) = \sqrt{\left(\frac{2}{3}(x^2 + 1)\right)^2 + 1} \quad (x \in \mathbb{Z}).$$

Thus

$$[h \circ (g \circ f)](x) = [(h \circ g) \circ f](x) \quad (x \in \mathbb{Z})$$

from which

$$h \circ (g \circ f) = (h \circ g) \circ f.$$

Theorem 7 Associativity of Circle-Composition

Let A, B, C, D be sets and let $f, g,$ and h be mappings such that
$$f : A \to B, \; g : B \to C, \; h : C \to D.$$

Then
$$h \circ (g \circ f) = (h \circ g) \circ f.$$

Proof

The proof simply entails repeated and careful application of Definition 13. Let $a \in A$. Then

$$[h \circ (g \circ f)](a) = h((g \circ f)(a))$$
$$= h(g(f(a)))$$

and

$$[(h \circ g) \circ f](a) = (h \circ g)(f(a))$$
$$= h(g(f(a))).$$

Thus

$$[h \circ (g \circ f)](a) = [(h \circ g) \circ f](a)$$

and since a is arbitrary

$$h \circ (g \circ f) = (h \circ g) \circ f.$$ □

We may now write, unambiguously, $h \circ g \circ f$ to mean either of the compositions $h \circ (g \circ f)$ or $(h \circ g) \circ f$.

Remark

Mappings may appear to be necessary but somewhat humdrum objects of study. However particular mappings may give rise to curious problems. Consider, for example, the mapping $f : \mathbf{N} \to \mathbf{N}$ given by

$$f(n) = \begin{cases} \frac{1}{2}n & (n \text{ even}) \\ \frac{1}{2}(3n + 1) & (n \text{ odd}). \end{cases}$$

Thus $f(26) = 13$, $f(25) = \frac{1}{2}(75 + 1) = 38$. Suppose we iterate this mapping several times and, by way of illustration, we follow the effect of the iterations on the number 10. Then

$$f(10) = 5,$$

$$(f \circ f)(10) = f(5) = 8,$$

$$(f \circ f \circ f)(10) = f(8) = 4,$$

$$(f \circ f \circ f \circ f)(10) = f(4) = 2,$$

$$(f \circ f \circ f \circ f \circ f)(10) = 1.$$

After five iterations on 10 the number 1 appears. The reader may care to verify that commencing with 65 there are 19 iterations before 1 first appears. Other integers may be tried at random as test cases, but the reader is advised to cultivate patience as the number of iterations before 1 first appears may be quite large.

The conjecture, that for any $n \in \mathbf{N}$ and for sufficiently many iterations the number 1 always appears, remains to be proved or disproved.

Exercises 1.3

1. Let A, B, C be the sets $\{a, b, c\}$, $\{p, q, r\}$, $\{0, 1\}$ respectively. Let the mappings $f : A \rightarrow B$ and $g : B \rightarrow C$ be defined by

$$f(a) = p, f(b) = f(c) = q, g(p) = 0, g(q) = 1, g(r) = 0.$$

Evaluate $(g \circ f)(a), (g \circ f)(b), (g \circ f)(c)$.

2. Let A, B, C, D be sets and let p, q, r, s be mappings such that

$$p : A \rightarrow B, q : B \rightarrow C, r : A \rightarrow B, s : D \rightarrow C.$$

Which of the following are defined: $q \circ p, p \circ q, p \circ r, p \circ s, r \circ s, q \circ s$?

3. Let f, g, h be mappings of $\mathbb{Z}$ into $\mathbb{Z}$ given by

$$f(n) = 2n, \; g(n) = 3n + 5, \; h(n) = -6n \quad (n \in \mathbb{Z}).$$

Which of the following pairs of mappings are equal: $f \circ g, g \circ f$; $f \circ h$, $h \circ f$; $g \circ h, h \circ g$?

4. Let A, B and C be sets and let $f : A \rightarrow B$ and $g : B \rightarrow C$ be mappings.

   (i)   If f and g are injective prove that $g \circ f$ is injective.

   (ii) If $g \circ f$ is injective, are g and f injective?

   (iii) If $g \circ f$ is surjective, are g and f surjective?

5. Let X, Y be sets and let $f : X \rightarrow Y$ be a mapping. Let A, B be subsets of X.

   (i)   If $A \subseteq B$ prove that $f(A) \subseteq f(B)$.

   (ii) Prove that $f(A \cup B) = f(A) \cup f(B)$.

   (iii) Prove that $f(A \cap B) \subseteq f(A) \cap f(B)$.

6. Let A be the subset of $\mathbb{Z}$ consisting of the even integers and let B be the subset of $\mathbb{Z}$ consisting of the odd integers. Let a mapping $f : \mathbb{Z} \rightarrow \mathbb{Z}$ be defined by

$$f(n) = n^2 - 3n + 5 \quad (n \in \mathbb{Z}).$$

Prove that $3 \in f(A) \cap f(B)$ and so deduce that $f(A \cap B) \neq f(A) \cap f(B)$.

1.4 Equivalence Relations

In the context of humans, animals and plants, the notion of relationship is all pervasive; two individuals are regarded as 'related' if they share a common parent, grandparent or even a more distant ancestor. Thus cousins are said to be related but we would not normally regard a cat and a dog as being related. In these illustrations we encounter the idea of a universal set, whether composed of humans or of domestic animals, and of the relationship which members or elements of the set may bear to one another under some particular criterion. Since the underlying concept of relationship is very useful in mathematics we discuss a more mathematical example before proceeding to a precise definition.

Example 24

We consider a set A and a relationship which exists between some pairs of elements of A. Let X be a set and let $f : A \to X$ be a mapping. We shall say that elements a and b of A are 'related' if $f(a) = f(b)$. By this definition any two elements of A may or may not be related.

For convenience we write for $a, b \in A$, $a \sim b$ if and only if a, b are related; the presence of the symbol $\sim$ (pronounced 'twiddles') denoting that a and b are related.

We note three obvious facts:

1. For all $a \in A$, a is related to a since $f(a) = f(a)$ implies $a \sim a$ (for all $a \in A$).
2. If $a, b \in A$ are such that a is related to b, then b is related to a, since $a \sim b$ implies $f(a) = f(b)$ and so $f(b) = f(a)$ and $b \sim a$.
3. If $a, b, c \in A$ are such that a is related to b and b is related to c, then a is related to c since $a \sim b$ implies $f(a) = f(b)$ and $b \sim c$ implies $f(b) = f(c)$ from which $f(a) = f(b) = f(c)$ and so $a \sim c$.

The definition we seek basically repeats the conclusions of this example.

Definition 14

Let A be a set. A relation, denoted by $\sim$, is defined between some pairs of elements of A subject to the following (named) conditions.

1. For all $a \in A$, $a \sim a$ (reflexivity).
2. If $a, b \in A$ are such that $a \sim b$ then also $b \sim a$ (symmetry).
3. If $a, b, c \in A$ are such that $a \sim b$ and $b \sim c$ then also $a \sim c$ (transitivity).

A relation $\sim$ on A which is reflexive, symmetric and transitive as above, is called an **equivalence relation**.

Symbols R, ρ, σ are frequently used to denote equivalence relations.

We note in passing that the relation of the previous example is an equivalence relation.

Examples 25

1. On any set A the relation of equality is an equivalence relation since we have $a = a$ for all $a \in A$, we certainly have $a = b$ implies $b = a(a, b \in A)$ and if $a = b$ and $b = c(a, b, c \in A)$ then $a = c$.

   An equivalence relation between the elements of a set generalizes the notion of equality for elements of the set.

2. On $\mathbf{Z}$ a relation $\sim$ is defined by $a \sim b$ if and only if a, b are both even or a, b are both odd.

   Thus $4 \sim 4$ and $5 \sim 5$ but we do not have $4 \sim 5$.

   Now, in general, $a \sim a$ and certainly $a \sim b$ implies $b \sim a$ $(a, b \in \mathbf{Z})$. Suppose now $a \sim b$ and $b \sim c$ $(a, b, c \in \mathbf{Z})$. Then, if b is even, a and c must also be even and $a \sim c$, whereas if b is odd, a and c must also be odd and $a \sim c$. Thus $\sim$ is an equivalence relation on $\mathbf{Z}$.

   Notice that we could have defined this equivalence relation by means of a mapping $f : \mathbf{Z} \to \{0, 1\}$ for which

   $$f(n) = \begin{cases} 0 & (n \text{ even}) \\ 1 & (n \text{ odd}) \end{cases} \quad (n \in \mathbf{Z}).$$

3. Let S be the set of all words in a given English dictionary. Define two words to be related by ρ if they begin with the same letter and end with the same letter. Thus we have 'atom ρ alum', 'pleasure ρ perseverance' and 'a ρ aroma'. Then ρ is easily seen to be an equivalence relation on S.

4. Let S be the set of points composing the circumference of a given circle. For P, $Q \in S$ define a relation σ by $P \sigma Q$ if and only if there is a diameter on which P, Q are points. Thus if $P \sigma Q$ then either P, Q coincide or P, Q are at opposite ends of a diameter. Simple geometrical considerations show that σ is an equivalence relation.

5. Let X be the set $\mathbf{Q} \times \mathbf{Q}$ of ordered pairs of rational numbers. For $a = (a_1, a_2) \in \mathbf{Q} \times \mathbf{Q}$ and $b = (b_1, b_2) \in \mathbf{Q} \times \mathbf{Q}$ we define a relation $\sim$ by

   $$a \sim b \text{ if and only if } a_1 + a_2 = b_1 + b_2.$$

   We claim that $\sim$ is an equivalence relation on X. Certainly $\sim$ is reflexive.

If for some $a, b \in X$ we have $a \sim b$ then $a = (a_1, a_2)$, $b = (b_1, b_2)$, where $a_i, b_i \in \mathbb{Q}$ $(i = 1, 2)$ and $a_1 + a_2 = b_1 + b_2$. But then $b_1 + b_2 = a_1 + a_2$ and so $b \sim a$ and $\sim$ is symmetric.

Suppose now for some $a, b, c \in X$ we have $a \sim b$ and $b \sim c$. Then writing $a = (a_1, a_2)$, $b = (b_1, b_2)$, $c = (c_1, c_2)$ we have $a_1 + a_2 = b_1 + b_2 = c_1 + c_2$ and so $a \sim c$ and $\sim$ is transitive. Hence $\sim$ is an equivalence relation.

6. A relation, ρ, is defined on $\mathbb{Q}$ by letting, for $a, b \in \mathbb{Q}$, $a\rho b$ if and only if we have $a - b \in \mathbb{Z}$. By this relation we have

$$\frac{43}{19} \rho \frac{5}{19} \quad \text{as} \quad \frac{43}{19} - \frac{5}{19} = \frac{38}{19} = 2 \in \mathbb{Z}$$

but

$$\frac{1}{2} \rho \frac{1}{3} \text{ is false as } \frac{1}{2} - \frac{1}{3} = \frac{5}{6} \notin \mathbb{Z}.$$

We may prove that ρ is an equivalence relation as follows:

(i) For all $a \in \mathbb{Q}$, $a - a = 0 \in \mathbb{Z}$ and so $a \sim a$ (reflexivity).

(ii) If for some $a, b \in \mathbb{Q}$, $a \sim b$ then $a - b \in \mathbb{Z}$ which immediately implies that $b - a = -(a - b) \in \mathbb{Z}$ and so $b \sim a$ (symmetry).

(iii) If for some $a, b, c \in \mathbb{Q}$, $a \sim b$ and $b \sim c$ then evidently $a - b \in \mathbb{Z}$ and $b - c \in \mathbb{Z}$ from which $a - c = (a - b) + (b - c) \in \mathbb{Z}$ and so $a \sim c$ (transitivity).

It is sometimes important, in regard to a set of conditions specifying a mathematical entity such as an equivalence relation, to know whether the stipulated conditions are themselves independent or not. We show that the three conditions of reflexivity, symmetry and transitivity for an equivalence relation are independent by exhibiting three examples in each of which two of the conditions are satisfied but not the remaining condition.

Examples 26

1. Let a relation $\sim$ be defined on $\mathbb{Z}$ by $a \sim b$ if and only if $ab \neq 0$ $(a, b \in \mathbb{Z})$. Then the relation is symmetric since $a \sim b$ implies $ab \neq 0$ and so $ba \neq 0$. Thus $b \sim a$ $(a, b \in \mathbb{Z})$.

The relation is also transitive since $a \sim b$, $b \sim c$ implies $ab \neq 0$, $bc \neq 0$. Hence $a \neq 0$, $b \neq 0$, $c \neq 0$ and thus $ac \neq 0$ giving $a \sim c$.

On the other hand the relation is not reflexive since we do not have $a \sim a$ for all $a \in \mathbb{Z}$, in particular we do not have $0 \sim 0$.

2. The relation of inequality, $\leq$, on $\mathbb{R}$ is reflexive and transitive since we have, for all $a \in \mathbb{R}$, $a \leq a$ and if $a \leq b$, $b \leq c$ then $a \leq c$ $(a, b, c \in \mathbb{R})$. We cannot infer from $a \leq b$ $(a, b \in \mathbb{R})$ that $b \leq a$ and so this relation is not symmetric.

3. Let a relation $\sim$ be defined on $\mathbb{Z}$ by $a \sim b$ if and only if 2 divides $a - b$ or 3 divides $a - b$. Such a relation is not transitive since, for example,

$7 \sim 5$ since $7 - 5 = 2$

and $5 \sim 2$ since $5 - 2 = 3$

but $7 \sim 2$ is false since $7 - 2 = 5$ which is divisible by neither 2 nor 3.

But the relation is reflexive since for all $a \in Z$, $a - a = 0$ and so $a \sim a$ and the relation is symmetric since $a \sim b$ $(a, b \in \mathbb{Z})$ implies that $a - b$, and thus $b - a$, is divisible by 2 or by 3 and so $b \sim a$.

We may consider the totality of the inhabitants of a given town as being distributed into the families which dwell in the town. Any two members of the same family are somehow related but no two members of different families are related. An inhabitant of the town belongs to only one family and so identifies uniquely the family to which he or she belongs. In a similar way a set with an equivalence relation will be seen to be the union of disjoint subsets called equivalence classes, each equivalence class consisting (like a family) of all elements which are related to one another and each equivalence class will be uniquely identifiable by any element belonging to it.

We make these ideas more precise.

Definition 15

Let S be a non-empty set with an equivalence relation $\sim$. Let $a \in S$. The subset S_a of S given by

$$S_a = \{x \in S : x \sim a\}$$

is said to be the **equivalence class** determined by, or containing, a (note that as $a \sim a$, $a \in S_a$).

We now prove a result which we shall have frequent occasion to use.

Theorem 8

In the notation of Definition 15 the following hold:

1. $S = \bigcup_{a \in S} S_a$.
2. $S_a = S_b$ if and only if $a \sim b$ $(a, b \in S)$.
3. Either $S_a = S_b$ or $S_a \cap S_b = \emptyset$ $(a, b \in S)$.

Proof

1. Since $a \in S_a$ it follows that $S \subseteq \bigcup_{a \in S} S_a$ and so $S = \bigcup_{a \in S} S_a$.
2. If $S_a = S_b$ then $a \in S_b$ and so $a \sim b$. Conversely suppose $a \sim b$. We prove that $S_a \subseteq S_b$. Let $x \in S_a$. Then $x \sim a$ and $a \sim b$ implies that $x \sim b$ and so $x \in S_b$. Thus $S_a \subseteq S_b$. Since, by symmetry, $b \sim a$ we also have $S_b \subseteq S_a$ and so $S_a = S_b$.
3. Suppose $S_a \cap S_b \neq \emptyset$. We show that $S_a = S_b$. Let $c \in S_a \cap S_b$ and so $c \sim a$ and $c \sim b$. But then $a \sim c$ and $c \sim b$ which implies, by transitivity, that $a \sim b$. By (ii) we conclude that $S_a = S_b$. □

It is a consequence of Theorem 8 that from the union $\bigcup_{a \in S} S_a$, on eliminating equivalence classes that formally coincide, we may write

$$S = \bigcup_{a \in T} S_a$$

where T is a subset of S such that S_a $(a \in T)$ are distinct, and $S_a \cap S_b = \emptyset$ $(a, b \in T, a \neq b)$. The non-empty set S is thereby expressed as a disjoint union of distinct equivalence classes.

Examples 27 (Continued from Examples 25)

We examine the examples immediately following Definition 14 to determine the appropriate equivalence classes.

1. With equality as the equivalence relation on a set each equivalence class consists of a single element.
2. There are two equivalence classes, namely the subset of $\mathbf{Z}$ consisting of the even integers and the subset of $\mathbf{Z}$ consisting of the odd integers.
3. The English alphabet has 26 letters: $a, b, c, \ldots, x, y, z$. Each word begins with, and ends with, one of these letters. We now define $26^2 = 676$ subsets each defined for any pair of letters of the alphabet as follows:

Let S_{aa} be the subset of all words beginning with a and ending with a.

Let S_{ab} be the subset of all words beginning with a and ending with b.

...   .

Let S_{zz} be the subset of all words beginning with z and ending with z.

The reader will immediately observe that some of the subsets $S_{aa}, S_{ab}, \dots, S_{zz}$ are empty ($S_{zz} = \varnothing$ but, perhaps surprisingly, $S_{az} \neq \varnothing$). Any one of these subsets, if non-empty, is an equivalence class. S is the union of these subsets and, on omitting empty subsets, becomes a disjoint union of equivalence classes. (We leave it to the aspiring lexicographer to determine which subsets are non-empty.)

4. Every diameter determines an equivalence class consisting of the two end-points of the diameter.

5. For every $q \in \mathbb{Q}$ there is an equivalence class given by

$$X_q = \{(a_1, a_2) \in Q \times Q | a_1 + a_2 = q\}.$$

Each equivalence class is of this form.

We have seen that if a set admits an equivalence relation then that set is a disjoint union of particular subsets called equivalence classes. Conversely if a set is a disjoint union of subsets then we shall show that, correspondingly, an equivalence relation may be defined on the set. We are led to make the following definition.

Definition 16

Let S be a non-empty set. Let $\{S_\lambda : \lambda \in \Lambda\}$ be a collection, indexed by an index-set Λ, of non-empty subsets of S such that

1. $S_\lambda \cap S_\mu = \varnothing \, (\lambda, \mu \in \Lambda, \lambda \neq \mu)$ and
2. $\bigcup_{\lambda \in \Lambda} S_\lambda = S$.

The collection of subsets is said to form a **partition** of S.

Suppose we have a non-empty set S with a partition as above. For $a, b \in S$ define a relation $\sim$ by letting $a \sim b$ if and only if a, b belong to the same subset S_λ (say) of the partition. Thus any two elements in an S_λ are related but no element in S_λ is related to an element in $S_\mu \, (\lambda \neq \mu)$. Then it may readily be verified that $\sim$ is an equivalence relation on S and that the equivalence classes are the subsets $S_\lambda \, (\lambda \in \Lambda)$. A partition therefore induces an equivalence relation and an equivalence relation induces a partition, subsets of the partition forming the equivalence classes of the equivalence relation.

Example 28

Let $S = \{a, b, c, d, e, f, g\}$. Let $S_1 = \{a, b, d\}$, $S_2 = \{c, g\}$, $S_3 = \{e, f\}$. Then the subsets S_1, S_2, S_3 form a partition $\{S_1, S_2, S_3\}$ of S. We define the corresponding equivalence relation $\sim$ on S as follows:

$$
\begin{aligned}
& a \sim a,\, a \sim b,\, a \sim d, \quad c \sim c, \quad e \sim e, \\
& b \sim a,\, b \sim b,\, b \sim d, \quad c \sim g, \quad e \sim f, \\
& d \sim a,\, d \sim b,\, d \sim d, \quad g \sim c, \quad f \sim e, \\
& \qquad\qquad\qquad\qquad\quad g \sim g, \quad f \sim f.
\end{aligned}
$$

Exercises 1.4

1. A relation ρ is defined on $\mathbb{R}$ by $a \rho b$ if and only if $a^2 = b^2$ $(a, b \in \mathbb{R})$. Prove that ρ is an equivalence relation on $\mathbb{R}$. Identify the equivalence classes.

2. A relation R is defined on $\mathbb{Z}$ by $a R b$ if and only if $a - b$ is even $(a, b \in \mathbb{Z})$. Prove that R is an equivalence relation on $\mathbb{Z}$. Identify the equivalence classes.

3. A relation τ is defined on $\mathbb{Z}$ by $a \tau b$ if and only if $a - b$ is divisible by 6 $(a, b \in \mathbb{Z})$. Prove that τ is an equivalence relation and determine the equivalence classes.

4. What are the equivalence classes in Examples 25, no. 6 immediately following Definition 14?

5. Let Oxy be the two-dimensional coordinate plane. Let P, Q be points of Oxy.

   (i) P, Q are said to be related by σ if the points O, P, Q are collinear. Is σ an equivalence relation?

   (ii) P, Q are said to be related by τ if there exists a rotation about O which sends P into Q. Is τ an equivalence relation?

   Sketch, if possible, the equivalence classes.

6. How many distinct equivalence relations may be defined on a set of one, two, three or four elements?

7. A relation ρ is defined on $\mathbb{Z}$ by $a \rho b$ if and only if $a - b$ is divisible either by 5 or by 7 $(a, b \in \mathbb{Z})$. Is ρ an equivalence relation?

8. Let $S = \{a, b, c, x, y, z\}$. Two relations ρ, σ are defined on S as follows:

   $a \rho a,\ b \rho b,\ c \rho c,\ x \rho x,\ y \rho y,\ z \rho z,\ a \rho b,\ b \rho a,\ a \rho c,\ c \rho a,\ b \rho c,\ c \rho b;\ a \sigma a,\ b \sigma b,\ c \sigma c,\ x \sigma x,\ y \sigma y,\ z \sigma z,\ a \sigma x,\ x \sigma a,\ b \sigma y,\ y \sigma b,\ c \sigma z,\ z \sigma c,\ a \sigma y,\ y \sigma a.$

   Do either ρ or σ define an equivalence relation on S?

9. Let S be a set on which a relation $\sim$ is defined. The relation $\sim$ is symmetric and transitive. Loose thinking would suggest that $a \sim b$ ($a, b \in S$) implies $b \sim a$ (by symmetry) and so $a \sim a$ (by transitivity) thus giving reflexivity, and consequently $\sim$ would be an equivalence relation. What is wrong with this loose thinking?

1.5 Well-ordering and Induction

The elements of the set N, that is the set of strictly positive integers, may be written down in ascending order with a repeated inequality sign, thus:

$$1 < 2 < 3 < 4 < \ldots$$

Let S be a non-empty subset of N. Let $N \in$ N. From the integers $1, 2, \ldots, N$ we may select that one which is least amongst these integers and which also belongs to S. Intuitively therefore we have the following principle which we shall accept as axiomatic.

Principle of Well-ordering in N

Every non-empty subset S of N has a least integer in S.

It frequently happens that we have some assertion, proposition or statement $P(n)$ which depends on the particular integer n. The proposition may itself be true or false. We give three examples.

Examples 29

1. $1 + 2 + \ldots + n = \dfrac{n}{2}(n + 1)$ $(n \in \text{N})$.

2. $2n + 1 \le 2^n$ $(n \in \text{N})$.

3. $n^2 \le 2^n$ $(n \in \text{N})$.

In each of these examples we have a statement that depends on n. We are not asserting the truth or falsity of the statement. Naturally, however, we wish to know whether the particular statement is true for all $n \in$ N or, possibly, for all

$n \in \mathbf{N}$ greater than some fixed integer. We are led to the so-called Principle of Induction and to an obvious and convenient variant of this principle. We may derive the Principle of Induction from the Principle of Well-ordering but both, for present purposes, may be regarded as simply axiomatic or, indeed, as 'obvious'.

Principle of Induction

Let $P(n)$ be a proposition depending on the integer n. Suppose that

1. $P(1)$ is true and
2. if $P(k)$ is true then $P(k+1)$ is true (induction assumption).

Then $P(n)$ is true for all $n \in \mathbf{N}$.

Proof (may be omitted)

Let S be the subset of $\mathbf{N}$ of those integers n for which $P(n)$ is true. Then certainly $1 \in S$ and so $S \neq \emptyset$. Let $X = \mathbf{N} \setminus S$. We wish to show that $X = \emptyset$ and then $S = \mathbf{N}$. For the sake of argument suppose $X \neq \emptyset$. We apply the Principle of Well-ordering to X. Let N be the least integer in X. Now $N \neq 1$ since $1 \in S$ and so $N > 1$. Since N is the least integer in X, $N - 1$ is an integer not in X and so $N - 1 \in S$. But then $P(N-1)$ is true and so, by hypothesis 2, $P(N)$ is also true and $N \in S$. But this is a contradiction as $X \cap S = \emptyset$. Hence $X = \emptyset$ and $S = \mathbf{N}$. □

The Principle of Induction is often used in a modified version which we may also deem to be axiomatic.

Principle of Induction (Modified Version)

Let $P(n)$ be a proposition depending on the integer n. Suppose that

1. $P(1)$ is true and
2. if for each $m \leq k$, $P(m)$ is true, then $P(k+1)$ is true (induction assumption).

Then $P(n)$ is true for all $n \in \mathbf{N}$.

It sometimes happens that in either version of the Principle of Induction statement 1 that '$P(1)$ is true' is replaced by 'For some $N \in \mathbf{N}$, $P(N)$ is true' with corresponding changes to statement 2. In this circumstance the conclusion is that $P(n)$ is true for all $n \in \mathbf{N}$, $n \geq N$.

Let us consider further examples:

Examples 30

1. $P(n)$ is the statement

$$1 + 2 + \ldots + n = \frac{n}{2}(n+1) \quad (n \in \mathbb{N}).$$

Certainly $P(1)$ is true since

$$1 = \frac{1}{2}(1+1).$$

If we now make the induction assumption that $P(k)$ is true, then we suppose that

$$1 + 2 + \ldots + k = \frac{k}{2}(k+1).$$

But this implies that

$$1 + 2 + \ldots + k + (k+1) = \frac{k}{2}(k+1) + (k+1)$$

$$= \frac{(k+1)}{2}(k+2)$$

$$= \frac{(k+1)}{2}[(k+1)+1]$$

and so we may assert that $P(k)$ implies $P(k+1)$. Hence we have for all $n \in \mathbb{N}$

$$1 + 2 + \ldots + n = \frac{n}{2}(n+1).$$

2. $P(n)$ is the statement that

$$2n + 1 \leq 2^n \quad (n \in \mathbb{N}).$$

Now $P(1)$ and $P(2)$ are, in fact, false since

$$2(1) + 1 > 2^1 \text{ and } 2(2) + 1 > 2^2.$$

However, $P(3)$ is true since

$$2(3) + 1 = 7 \leq 2^3.$$

Let us suppose that $P(k)$ is true for all $k \geq 3$. Then we suppose

$$2k + 1 \leq 2^k \quad (\text{if } k \geq 3).$$

But this implies that

$$2(k+1) + 1 = 2k + 3 = 2k + 1 + 2 \leq 2^k + 2 \leq 2^k + 2^k = 2^{k+1} \quad (k \geq 3)$$

and so $P(k+1)$ is true. Hence we conclude that for all $n \in \mathbb{N}$, $n \geq 3$

$$2n + 1 \leq 2^n.$$

3. $P(n)$ is the statement that

$$n^2 \leq 2^n \quad (n \in \mathbb{N}).$$

$P(1)$ and $P(2)$ are true since

$$1^2 = 1 \leq 2^1, 2^2 = 4 \leq 2^2$$

but $P(3)$ is false since

$$3^2 = 9 > 2^3.$$

However, $P(4)$ is true since

$$4^2 \leq 2^4.$$

We make the induction assumption that $P(k)$ is true for all $k \geq 4$. But then

$$k^2 \leq 2^k \quad (\text{if } k \geq 4)$$

implies that

$$(k+1)^2 = k^2 + 2k + 1 \leq 2^k + 2k + 1 \leq 2^k + 2^k = 2^{k+1} \quad (\text{by 2 above})$$

and so $P(k+1)$ is true. Hence we conclude that for all $n \in \mathbb{N}$, $n \geq 4$,

$$n^2 \leq 2^n.$$

When we attempt to apply induction as above we must be on guard in regard to two aspects. We must ensure that we do not go blindly on from the truth of $P(1)$ (say) and try to prove $P(k+1)$ from $P(k)$ when in fact $P(2)$ or $P(3)$ or whatever may be false (as in the Example above). The second aspect which may cause trouble is an invalid use of a particular induction assumption. We illustrate this aspect in the next example.

Example 31

By induction we shall 'prove' that, given any set of billiard balls on a billiard table, the balls are necessarily of the same colour.

Our statement $P(n)$ is that if n balls or fewer are on a billiard table they are of the same colour. Certainly if there is only one ball there is only one colour and $P(1)$ is true. We now try to prove, from the assumption that if we have k balls or fewer then they are of the same colour, that we may deduce that if we have $k+1$ balls or fewer then they are of the same colour.

Given a set S of $k+1$ balls we may select two distinct subsets A, B of S of k balls each. Then $S = A \cup B$ and $A \cap B \neq \emptyset$. Now by the induction hypothesis the balls in A are of one colour and the balls in B are of one colour. But A and B have at least one ball in common and so the balls in A and B must be of a single colour. Hence, as $A \cup B = S$, the balls in S are all of the same colour and we have completed the induction step of the argument, namely $P(k)$ implies

$P(k+1)$. Hence $P(n)$ is true for all $n \in \mathbb{N}$ and the balls on a billiard table are all of the same colour.

The purported conclusion is clearly false – so what is wrong in our arguments? We were careless in supposing that $P(2)$ is a consequence of $P(1)$. A moment's reflection should reveal that $P(2)$ is false.

This example should serve as a warning that, in applying induction, care must be taken to ensure that the conditions are properly met.

We give some further examples in the use of induction. In these examples we shall omit specific reference to a $P(n)$ but we use the convention that

$$\sum_{r=1}^{n} a_r = a_1 + a_2 + \ldots + a_n.$$

Examples 32

1. Prove that

$$\sum_{r=1}^{n} r^2 = \frac{n(n+1)(2n+1)}{6} \quad (n \in \mathbb{N}).$$

Certainly for $n = 1$ we have

$$1^2 = 1 = \frac{1(1+1)(2+1)}{6}$$

and so the statement holds for $n = 1$. We now suppose that

$$\sum_{r=1}^{k} r^2 = 1^2 + 2^2 + \ldots + k^2 = \frac{k(k+1)(2k+1)}{6}.$$

Then

$$\sum_{r=1}^{k+1} r^2 = 1^2 + 2^2 + \ldots + (k+1)^2 = (1^2 + 2^2 + \ldots + k^2) + (k+1)^2$$

$$= \frac{k(k+1)(2k+1)}{6} + (k+1)^2$$

$$= \frac{(k+1)}{6} [k(2k+1) + 6(k+1)]$$

$$= \frac{(k+1)}{6} [2k^2 + 7k + 6]$$

$$= \frac{(k+1)}{6} (k+2)(2k+3)$$

$$= \frac{(k+1)}{6} [(k+1) + 1][2(k+1) + 1].$$

This completes the induction argument and so the statement is true for all $n \in \mathbb{N}$.

2. Prove that

$$\sum_{r=1}^{n} \frac{1}{r^2} = \frac{1}{1^2} + \frac{1}{2^2} + \ldots + \frac{1}{n^2} \leq 2 - \frac{1}{n} \quad (n \in \mathbb{N}).$$

Certainly for $n = 1$ we have

$$\frac{1}{1} = 1 = 2 - \frac{1}{1}.$$

Suppose

$$\sum_{r=1}^{k} \frac{1}{r^2} = \frac{1}{1^2} + \frac{1}{2^2} + \ldots + \frac{1}{k^2} \leq 2 - \frac{1}{k}.$$

Then

$$\sum_{r=1}^{k+1} \frac{1}{r^2} = \frac{1}{1^2} + \frac{1}{2^2} + \ldots + \frac{1}{k^2} + \frac{1}{(k+1)^2} \leq 2 - \frac{1}{k} + \frac{1}{(k+1)^2}.$$

If we now focus our attention on the desired outcome of the induction argument we realize that we would like to show that

$$2 - \frac{1}{k} + \frac{1}{(k+1)^2} \leq 2 - \frac{1}{(k+1)}.$$

But this inequality is true since

$$\left(2 - \frac{1}{k+1}\right) - \left(2 - \frac{1}{k} + \frac{1}{(k+1)^2}\right) = \frac{1}{k} - \frac{1}{k+1} - \frac{1}{(k+1)^2}$$

$$= \frac{1}{k(k+1)} - \frac{1}{(k+1)^2} \geq 0.$$

Hence

$$\frac{1}{1^2} + \frac{1}{2^2} + \ldots + \frac{1}{(k+1)^2} \leq 2 - \frac{1}{k+1}$$

and the statement is true for all $n \in \mathbb{N}$.

Exercises 1.5

Establish the following eleven statements by induction:

1. $\displaystyle\sum_{r=1}^{n} r(r+1) = \frac{1}{3}n(n+1)(n+2) \quad (n \in \mathbb{N})$.

2. $\displaystyle\sum_{r=1}^{n} r^3 = \frac{1}{4}n^2(n+1)^2 \quad (n \in \mathbb{N})$.

3. $\displaystyle\sum_{r=1}^{n} a^{r-1} = 1 + a + \ldots + a^{n-1} = \begin{cases} \dfrac{1-a^n}{1-a} & (a \in R, a \neq 1) \\ n & (a=1) \quad (n \in \mathbb{N}). \end{cases}$

4. $\displaystyle\sum_{r=1}^{n} \frac{1}{r(r+1)} = \frac{n}{n+1} \quad (n \in \mathbb{N})$.

5. $\displaystyle\sum_{r=1}^{n} (5r-2) = \frac{n}{2}(5n+1) \quad (n \in \mathbb{N})$.

6. $\displaystyle\sum_{r=1}^{n} r2^r = 2 + (n-1)2^{n+1} \quad (n \in \mathbb{N})$.

7. $(1+x)^n \geq 1 + nx \quad (x \in \mathbb{R}, x \geq 0)$.

8. $3(2n+1) \leq 2^n \quad (n \in \mathbb{N}, n \geq 6)$.

9. $3n^2 \leq 2^n \quad (n \in \mathbb{N}, n \geq 8)$.

10. $2(\sqrt{n+1} - 1) \leq \displaystyle\sum_{r=1}^{n} \frac{1}{\sqrt{r}} \quad (n \in \mathbb{N})$.

11. The Fibonacci sequence $u_1, u_2, \ldots$ is defined inductively by

$$u_1 = 1, u_2 = 2, u_{n+1} = u_n + u_{n-1} \quad (n \geq 2).$$

Write down the first nine terms of the sequence and prove independently the two following results:

(i) $u_n \leq \left(\dfrac{7}{4}\right)^n \quad (n \in \mathbb{N})$.

(ii) $u_n = \dfrac{1}{\sqrt{5}}\left(\dfrac{1+\sqrt{5}}{2}\right)^{n+1} - \dfrac{1}{\sqrt{5}}\left(\dfrac{1-\sqrt{5}}{2}\right)^{n+1} \quad (n \in \mathbb{N})$.

The sequence was originally devised by Leonardo of Pisa, also called Fibonacci (c. 1170–1240), to model the procreative habits of amorous rabbits. Leonardo, through his text *Liber abbaci*, was influential in spreading the use of Hindu-Arabic numerals in Europe. We do well to remember the inestimable benefits these numerals have conferred and to acknowledge our intellectual debt to the civilizations from which they came. In the next chapter we shall exploit some of the advantages of the numerals.

1.6 Countable Sets

The sets $\{a, b, c, d\}$, $\{\alpha, \beta, \gamma, \delta\}$, $\{1, 2, 3, 4\}$ have no elements in common yet they share the property of being finite sets with the same number of elements. As we have seen, we may write the elements of a set X which has a finite number N of elements as

$$X = \{x_1, x_2, \ldots, x_N\},$$

in other words there is a bijection from $\{1, 2, \ldots, N\}$ to the set X. If a set X does not have a finite number of elements then X is infinite and there may or may not be a bijection from $\mathbf{N}$ to X. It is convenient to distinguish between 'levels' of infinity – as a porcine Napoleon might have decreed "All infinite sets are non-finite but some infinite sets are more infinite than others." (after *Animal Farm* by George Orwell, 1945). More precisely, we make the following definition to distinguish between those sets that admit a bijection from $\mathbf{N}$, or from a subset of $\mathbf{N}$, and those that do not.

Definition 17

A non-empty set X is said to be **countable** if the elements of X may be enumerated as a finite or infinite sequence of the form $x_1, x_2, x_3, \ldots$ If X is finite and has N elements then $X = \{x_n | n = 1, 2, \ldots, N\} = \{x_1, x_2, \ldots, x_N\}$ and if X is infinite then $X = \{x_n | n \in \mathbf{N}\} = \{x_1, x_2, \ldots\}$. Otherwise X is said to be **uncountable**.

Examples 33

1. The set $-\mathbf{N}$ of strictly negative integers given by $-\mathbf{N} = \{-n | n \in \mathbf{N}\}$ is countable since we have the sequence $-1, -2, -3, \ldots$ The bijection f from $\mathbf{N}$ to $-\mathbf{N}$ is given by $f(n) = -n$ $(n \in \mathbf{N})$.
2. The set $\mathbf{Z}$ of all integers is countable since we have the sequence $0, 1, -1, 2, -2, 3, -3, \ldots$ The bijection f from $\mathbf{N}$ to $\mathbf{Z}$ is given by

$$f(n) = \begin{cases} \dfrac{n}{2} & (n \text{ even}) \\ -\left(\dfrac{n-1}{2}\right) & (n \text{ odd}). \end{cases}$$

Since the elements of a countable set may be enumerated as a finite or infinite sequence it follows that the elements of any non-empty subset may also be enumerated as a sequence, namely as the subsequence of the given sequence obtained by omitting those elements not in the subset. Consequently we have the following result which we state for completeness.

Theorem 9

Any subset of a countable set is countable.

We shall prove that the set $\mathbb{Q}$ of rational numbers is countable but first we must investigate the countability of an array.

Remark

Let X be a set of elements x_{ij} having double suffices such that

$$X = \{x_{pq} \mid p, q \in \mathbb{N}\}.$$

We may write the elements down in the form of an infinite array similar to a matrix array:

$$
\begin{array}{lllll}
x_{11} & x_{12} & x_{13} & x_{14} & \cdots \\
x_{21} & x_{22} & x_{23} & x_{24} & \cdots \\
x_{31} & x_{32} & x_{33} & x_{34} & \cdots \\
x_{41} & x_{42} & x_{43} & x_{44} & \cdots \\
\cdots & \cdots & \cdots & \cdots & \cdots
\end{array}
$$

We begin the first stage of the enumeration of these elements by choosing x_{11}. For the second stage we choose those elements of which the sums of their suffices are 3 and we take them in the order of first suffix. At the third stage we choose those elements of which the sums of their suffices are 4 and we again take them in order of the first suffix. Continuing we obtain a sequence as follows:

$$x_{11}; x_{12}, x_{21}; x_{13}, x_{22}, x_{31}; x_{14}, x_{23}, x_{32}, x_{41}; \ldots$$

Each element of the array appears precisely once in this sequence and so X is countable.

Before drawing general conclusions let us determine explicitly the bijection f from $\mathbb{N}$ to X. Certainly we have $f(x_{11}) = 1, f(x_{12}) = 2, f(x_{21}) = 3, \ldots$ We require to determine the integer to which x_{pq} corresponds. Now x_{pq} is the pth element in the stage of choosing in which we commence with $a_{1,p+q-1}$.

The stages prior to this stage have involved successively $1, 2, 3, \ldots, p + q - 2$ elements. Thus x_{pq} is the element in the $[[1 + 2 + \ldots + (p + q - 2)] + p]$th place in the sequence of enumeration. But, from an earlier example, we have

$$f(x_{pq}) = [1 + 2 + \ldots + (p + q - 2)] + p$$

$$= \frac{1}{2}(p + q - 2)(p + q - 1) + p$$

$$= \frac{1}{2}[(p + q)^2 - p - 3q + 2].$$

This gives, for example,

$$f(x_{34}) = \frac{1}{2}[(3 + 4)^2 - 3 - 12 + 2] = 18$$

and we may verify directly that x_{34} is in the 18th place. We may now derive several useful conclusions.

Theorem 10

The union of a countable number of sets, each of which is itself countable, is countable.

Proof

Let $X_1, X_2, \ldots$ be the countable sets of the union.

Let X_1 have elements $x_{11}, x_{12}, \ldots$

Let X_2 have elements $x_{21}, x_{22}, \ldots$

$\ldots \qquad \ldots \qquad \ldots \qquad \ldots$

Let X_n have elements $x_{n1}, x_{n2}, \ldots$

$\ldots \qquad \ldots \qquad \ldots \qquad \ldots$

Thus the elements of $\bigcup_n X_n$ may be written in an array identical to that in the Remark above. We know that the array above is countable and so $\bigcup_n X_n$ is countable. □

Example 34

Since N, −N and $\{0\}$ are countable, so also is

$$\mathbf{Z} = -\mathbf{N} \cup \{0\} \cup \mathbf{N}.$$

Theorem 11

The Cartesian product of a finite number of countable sets is countable.

Proof

We prove the result first for the Cartesian product of two countable sets A, B.

Let A have elements $a_1, a_2, \ldots$ and let B have elements $b_1, b_2, \ldots$ Then the elements of $A \times B$ may be written in the array

$$(a_1, b_1), (a_1, b_2), (a_1, b_3), \ldots$$
$$(a_2, b_1), (a_2, b_2), (a_2, b_3), \ldots$$
$$\cdots \qquad \cdots \qquad \cdots$$

The array is countable and so is $A \times B$.

If now we have countable sets $X_1, X_2, \ldots, X_N$, then we have

$$X_1 \times X_2 \times \ldots \times X_N = (X_1 \times X_2 \times \ldots \times X_{N-1}) \times X_N$$

and a simple application of the Principle of Induction gives the final result. $\square$

Theorem 12

The set of rational numbers $\mathbb{Q}$ is countable.

Proof

Every strictly positive rational number is of the form $\frac{m}{n}$ for some $m, n \in \mathbb{N}$ and so may be identified with the ordered pair $(m, n) \in \mathbb{N} \times \mathbb{N}$. Thus $\frac{2}{3}$ may be identified with $(2, 3)$ and $\frac{4}{6}$ may be identified with $(4, 6)$; of course, purely as numbers, $\frac{2}{3} = \frac{4}{6}$ but formally $\frac{2}{3}$ and $\frac{4}{6}$ are distinct as the ordered pairs $(2, 3)$ and $(4, 6)$ are certainly distinct. Thus the set of strictly positive rational numbers $\mathbb{Q}^+$ is identified with $\mathbb{N} \times \mathbb{N}$ and so, by Theorem 11, $\mathbb{Q}^+$ is countable.

But letting $-\mathbb{Q}^+ = \{-q | q \in \mathbb{Q}^+\}$ we see that $-\mathbb{Q}^+$ is countable and so finally

$$\mathbb{Q} = -\mathbb{Q}^+ \cup \{0\} \cup \mathbb{Q}^+ \text{ is countable.} \qquad \square$$

Not all commonly occurring sets are countable. We state, but do not prove, the following result.

Theorem 13

The set of real numbers $\mathbb{R}$ is uncountable.

Nevertheless it would be useful to exhibit at least one example of uncountability as follows.

Example 35

Let A be the set of all mappings with domain N and range $\{0, 1\}$. We claim that A is uncountable. We shall argue by contradiction in supposing that we have included all such mappings in a sequence of enumeration and then, like a conjuror producing a rabbit out of a hat, we shall produce a mapping not previously in the sequence. This then will be our contradiction.

Suppose, therefore, we have enumerated all of the given mappings from N to $\{0, 1\}$ in the sequence $f_1, f_2, \ldots$ Then every mapping from N to $\{0, 1\}$ occurs as an f_r for some $r \in N$. We define $f : N \to \{0, 1\}$ as follows:

$$f(n) = \begin{cases} 0 & \text{if } f_n(n) = 1 \\ 1 & \text{if } f_n(n) = 0 \end{cases} \quad (n \in N).$$

Then for all $n \in N$, $f(n) \neq f_n(n)$ and so $f \neq f_n$ $(n \in N)$. But then f cannot have occurred in the sequence $f_1, f_2, \ldots$ Hence the supposition of countability is invalid and so A is uncountable.

It is not the purpose of this text to give a logically impeccable account of the theory of sets, fundamental as this may be for mathematics. The reader should be aware, however, that matters in the affairs of sets, as in the affairs of hearts, may not be as straightforward as they appear. As illustration we offer some examples upon which the reader may exercise his or her analytical skills.

Examples of paradoxes

1. Epimenides of Crete is supposed to have said "Cretans are always liars" with the meaning that Cretans tell only lies. Was Epimenides lying or not when he made his statement? (Epimenides, 6th century BC, is believed to be the 'prophet' of St Paul's Epistle to Timothy, Chapter I, verse 12).

2. In a certain village there is a barber who shaves all and only those persons in the village who do not shave themselves. Does the barber shave himself? (attributed to B. Russell, 1872–1970).

3. In a certain country every prefecture must have a prefect and no two may have the same prefect. A prefect may be non-resident in the prefecture. The government passes a law creating a special area for non-resident prefects and obliging all non-resident prefects to live there. Eventually there are so many non-resident prefects in this area that the government declares it to be a prefecture. Where does the prefect of this new prefecture reside?

Exercises 1.6

1. Prove that $\mathbf{N} \times \mathbf{N}, \mathbf{N} \times \mathbf{N} \times \mathbf{N}, \mathbf{Z} \times \mathbf{Z}, \mathbf{Z} \times \mathbf{Z} \times \mathbf{Z}$ are countable sets.

2. Prove that $\{a + b\sqrt{2} \,|\, a, b \in \mathbf{Z}\}$ is countable.

3. Let A, B be sets. Prove that A is countable if and only if $A \cap B$, $A \setminus B$ are countable sets.

4. Let $\mathbf{N}^{\mathbf{N}}$ denote the set of all mappings of $\mathbf{N}$ into $\mathbf{N}$. Prove that $\mathbf{N}^{\mathbf{N}}$ is uncountable.

5. Let the elements of the set X be denoted by x_{pq} where p is an element of $\mathbf{N}$ and q is an element of $\{1, 2, \ldots, N\}$ for some given N. Prove that X is countable.

2
The Integers

Numbers are encountered early in life and in many practical contexts. In our infancy we develop a feeling for the natural numbers by chanting "one, two, three, ...", "eins, zwei, drei, ...", "yî, èr, sân, ...", or their equivalent, in whatever may be our mother tongue. By childhood we have assimilated, without too much conscious effort, the elementary properties of the addition, subtraction, multiplication and division of the natural numbers; in this text we take these elementary properties for granted. In such an approach we are following the celebrated dictum of L. Kronecker (1823–91), namely, "Die ganze Zahl schuf der liebe Gott; alles übrige ist Menschenwerk" which we may render in English as "God created the integers; everything else is man's handiwork". (Quotation from Philosophie der Mathematik und Naturwissenschaft by H. Weyl, R. Oldenburg, München, 1928, Section 6, page 27).

2.1 Divisibility

Before considering concepts of division and divisibility we recall some earlier notation. We introduced the set $\mathbf{N}$ of natural numbers, $\mathbf{N} = \{1, 2, 3, \ldots\}$, the set $\mathbf{Z}$ of integers, $\mathbf{Z} = \{\ldots, -2, -1, 0, 1, 2, \ldots\}$ and the set $\mathbf{Q}$ of rational numbers $\mathbf{Q} = \{\frac{m}{n} \mid m, n \in \mathbf{Z}, n \neq 0\}$. (In passing we should remark that in some texts the symbol $\mathbf{N}$ denotes the set $\{0, 1, 2, \ldots\}$. The reader should confirm the significance of $\mathbf{N}$ in any text he or she happens to use.) We shall exploit properties of these sets.

Definition 1

Let a and b be integers with b being non-zero. Then we say that b **divides** a, or that b is a **divisor** of a, if there exists an integer c such that $a = bc$.

Notice that 0 is divisible by any integer b, $b \neq 0$, since $0 = b0$ and that any integer $a \neq 0$ has the so-called **trivial divisors** $\pm 1, \pm a$.

Examples 1

1. We have that 3 divides 6 since $6 = 3.2$ (where . denotes multiplication) but 4 does not divide 6 since there is no integer x such that $6 = 4x$. We have $6 = 4\frac{2}{3}$ but $\frac{3}{2} \notin \mathbf{Z}$.
2. We have that 36 divides $180 = 36.5$ but 24 does not divide 180.
3. 7 has the trivial divisors $\pm 1, \pm 7$. There are no other divisors.
4. 10 has the trivial divisors $\pm 1, \pm 10$. The remaining divisors are $\pm 2, \pm 5$.

Lemma 1

Let a, b and c be integers with b and c being non-zero. Let b divide a and c divide b. Then c divides a.

Proof

By assumption there exist integers d and e, such that $a = bd$, $b = ce$. Then
$$a = bd = (ce)d = c(ed).$$
Hence we conclude that c divides a. □

Examples 2

1. 25 divides 175 and 175 divides 700 and so, as in Lemma 1, 25 divides 700. (In fact $175 = 25.7, 700 = 175.4$ and $700 = 25.28$.)
2. The divisors of 18 (and also of -18) are $\pm 1, \pm 2, \pm 3, \pm 6, \pm 9, \pm 18$.
3. The divisors of 45 (and also of -45) are $\pm 1, \pm 3, \pm 5, \pm 9, \pm 15, \pm 45$.

As the repetition of $\pm$ signs becomes tedious we shall prove the following fairly obvious lemma.

Lemma 2

Let a and b be non-zero integers. Then the following are equivalent: (i) b divides a, (ii) b divides $-a$, (iii) $-b$ divides a, (iv) $-b$ divides $-a$.

Proof

If b divides a then $a = bc$ for some $c \in \mathbf{Z}$. Then $(-a) = b(-c)$, $a = (-b)(-c)$, $-a = (-b)c$ from which, respectively, b divides $-a$, $-b$ divides a, $-b$ divides $-a$. Thus (i) implies (ii), (iii) and (iv). The remaining implications are similarly established.   □

Lemma 2 ensures that, when we are seeking to test for the divisibility of one integer by another, we may suppose that both integers are strictly positive. In other words, to investigate divisibility in $\mathbf{Z}$ we may as well confine our attention to divisibility in $\mathbf{N}$ since we may attach $\pm$ as appropriate. We therefore wish to distinguish those integers in $\mathbf{N}$ which are in some sense (to be made precise later) 'irreducible' for division.

Definition 2

A strictly positive integer p, $p \neq 1$, is said to be a **prime number** or, simply, a **prime** if p has only trivial divisors. Thus if $n \in \mathbf{Z}$, $n \neq 0$ and n divides p then $n = \pm 1$ or $n = \pm p$. A strictly positive integer p, $p \neq 1$, which is not a prime is said to be **composite**.

The primes in ascending order are 2, 3, 5, 7, 11, 13, 17, 19, 23, 29, 31, 37, ...

Examples 3

1. Suppose we wish to find the next prime after 37. We run through the integers in ascending order. 38, 39, 40 are not primes since $38 = 2.19$, $39 = 3.13$, $40 = 2^3.5$. However 41 is not divisible by any $n \in \mathbf{N}$ except 1 and 41. Thus 41 is the next prime.

2. The number 178350 is composite and when expressed as a product of primes, or factorized as we say, is given by

$$178350 = 2.3.5^2.29.41.$$

To obtain this factorization we begin by observing that 10 divides 178350 and so $178350 = 17835.10$. Clearly 5 divides 17835 and $17835 = 3567.5$. We have disposed of obvious divisors such as 10 and 5. We now seek divisors of 3567.

Considering the primes in turn we find that $3567 = 1189.3$ and that the next prime dividing 1189 is 29 and $1189 = 41.29$. Thus finally

$$178350 = 17835.10 = 3567.5.10 = 1189.3.5.10 = 41.29.3.5.10$$

$$= 41.29.3.5.2.5 = 2.3.5^2.29.41.$$

We are familiar, as above, with the factorization of a given integer into the product of primes. In order to factorize a given integer we must determine the primes dividing the integer. Factorization involves division by arbitrary primes, whereas the definition of primeness (often called 'primality') of an integer relates to the divisibility of the integer. Once we have established certain properties of divisors we shall return to the primes.

Exercises 2.1

1. Write down the divisors of (i) 72, (ii) 84, (iii) 667.

2. Complete the proof of Lemma 2.

3. Factorize (i) 24, (ii) 93, (iii) 1225, (iv) 4913, (v) 14256.

2.2 Divisors

We are here mainly concerned with two given integers and with finding integers which divide them simultaneously. For this purpose we shall describe algorithms leading to the determination of such 'common divisors'. (The word 'algorithm' comes from the Latin 'algorismus' which is derived from the name of al-Khwarizmi, a 9th century Arab mathematician in Baghdad.) In mathematics, by the word 'algorithm', we understand a procedure which after a finite number of repetitions will yield some desired result.

Definition 3

Let a, b and c be non-zero integers. Then c is said to be a **common divisor** of a and b if c divides a and c divides b.

We may here usefully remark that if c divides a and b then c divides $a + b$ and $a - b$. This fact will be used frequently without further comment.

Example 4

The common divisors of -24 and 54 are $\pm1, \pm2, \pm3, \pm6$. We note that every common divisor divides ±6.

Definition 4

Let a and b be integers. Let a common divisor d have the property that, for any common divisor c of a and b, c divides d. Then d is called a **greatest common divisor** (G.C.D.).

If d' is a second greatest common divisor then certainly d' divides d and d divides d' and consequently $d' = \pm d$. That greatest common divisor which is positive is called **the greatest common divisor** and is denoted by (a, b). (The present meaning of this notation has to be distinguished from its usage as an ordered pair and from a common usage in analysis wherein (a, b) denotes the open interval $\{x \in R | a < x < b\}$. The context should make the meaning clear.)

Example 5

$$(4, 7) = (4, -7) = (-4, 7) = (-4, -7) = 1,$$
$$(6, 4) = (6, -4) = (-6, 4) = (-6, -4) = 2,$$
$$(36, 120) = 12, \quad (74, 111) = 37.$$

The greatest common divisor of two integers may be found by factorizing the integers into products of primes and then by comparing the factorizations. The drawback of this method is that it may be difficult to arrive at the appropriate factorizations; nevertheless the method may be used to find the greatest common divisor of small integers. We remark here that the method depends implicitly on the uniqueness of factorization of the integers, we shall consider this aspect more fully in the section on primes.

Examples 6

1. By the trial and error method as above we find that

$$616 = 2^3.7.11, \quad 1260 = 2^2.3^2.5.7.$$

Then

$$(616, 1260) = 2^2.7 = 28.$$

We observe that

$$616 = 28.22, \quad 1260 = 28.45,$$
$$(22, 45) = 1.$$

2. Similarly we have

$$1512 = 2^3.3^3.7, \quad 7056 = 2^4.3^2.7^2$$

and so

$$(1512, 7056) = 2^3.3^2.7 = 504.$$

We observe that

$$1512 = 504.3, \quad 7056 = 504.14, \quad (3, 14) = 1.$$

Exercise 2.2

1. By factorizing, find (a, b) for each pair of given integers a and b: (i) $a = 12$, $b = 15$; (ii) $a = 36$, $b = 60$; (iii) $a = 24$, $b = 35$; (iv) $a = 885$, $b = 1995$.

2.3 Division Algorithm

We now develop a more systematic approach to finding the G.C.D. of two integers. We begin with the so-called Division Algorithm of which we shall give two versions, the second generalizing the first.

Theorem 1 The Division Algorithm (First Version)

Let a and b be integers, b being non-zero. Then there exist positive, possibly zero, integers q and r satisfying the condition

$$a = bq + r \text{ where } 0 \leq r < b.$$

Proof

We notice first that we may dispose easily of the two cases when $a = b$ and $a < b$. For $a = b$ we have

$$a = b1 + 0 \quad (q = 1, r = 0)$$

and for $a < b$ we have

$$a = b0 + a \quad (q = 0, r = a).$$

These two cases also prove the result for $a = 1$. We shall apply the Principle of Induction and make the induction assumption that the result is true for $1, 2, \ldots, a - 1$ and try to obtain the result for a. Once we have established this conclusion we know that the result is true in general.

By the remarks above we need only consider the case of $b < a$. Then certainly $1 \le a - b$ and so by our remarks above (if $a - b \le b$) or by our induction assumption (if $b < a - b$) there exist integers q' and r such that

$$a - b = bq' + r, \quad 0 \le r < b,$$

from which

$$a = bq + r, \quad 0 \le r < b,$$

on letting $q = 1 + q'$. This completes the proof. □

At first sight the result above seems to be purely an 'existence proof' implying that for a given a and b there are appropriate q and r, but, like most existence proofs, no means by which we can calculate q and r are evident. In fact we do have a direct method, based on the above, of calculating the q (quotient) and the r (remainder), that of 'long division'. Formally

$$\frac{a}{b} = q + \frac{r}{b} \quad (0 \le r < b),$$

from which

$$a = bq + r \quad (0 \le r < b).$$

We illustrate this process below.

Examples 7

1. $a = 57, b = 19$. Evidently $q = 3, r = 0$.
2. $a = 527, b = 47$. By long division

$$
\begin{array}{r}
11 \\
47{\overline{)527}} \\
47 \\
\hline
57 \\
47 \\
\hline
10
\end{array}
$$

Thus $527 = 47.11 + 10$ and so $q = 11, r = 10$.

3. $a = 273921, b = 931$. By long division

$$
\begin{array}{r}
294 \\
931\overline{)273921} \\
1862 \\
\hline
8772 \\
8379 \\
\hline
3931 \\
3724 \\
\hline
207
\end{array}
$$

Thus

$$273921 = 931.294 + 207$$

and so

$$q = 294, \quad r = 207.$$

For the Division Algorithm (Extended Version) we introduce a convenient notation.

Definition 5

Let x be a real number. We define the **modulus**, or the **absolute value**, of x, denoted by $|x|$, as follows:

$$|x| = x \quad (x \geq 0), \qquad |x| = -x \quad (x < 0).$$

It is an immediate consequence of the definition that

$$|x| \geq 0 \text{ and } |x| = 0 \text{ if and only if } x = 0$$

and that

$$|x| = |y| \text{ if and only if } x = \pm y \quad (x, y \in \mathbb{R}).$$

Example 8

$$|7| = 7, \quad |-7| = -(-7) = 7.$$

Theorem 2 Division Algorithm (Extended Version)

Let a and b be integers, b being non-zero. Then there exist integers q and r satisfying the condition $a = bq + r$ where $0 \leq r < |b|$.

Proof

By the Division Algorithm (First Version) there exist $q, r \in \mathbb{N} \cup \{0\}$ such that

$$|a| = |b|q + r, \quad 0 \le r < |b|.$$

We consider several cases.

If $r = 0$ then $|a| = |b|q$ and so $a = \pm bq$ from which $a = b(\pm q)$ giving the result.

If $r > 0$ then we consider two cases. If $a \ge 0$ then $a = |b|q + r$ and so

$$a = bq + r \ (b > 0) \text{ and } a = (-b)q + r = b(-q) + r \ (b < 0),$$

of which only the second statement is wholly new. If $a < 0$ then $-a = |b|q + r$ and so $a = -|b|q - r$ which is almost in the desired form except that we require to have a positive remainder. We make a small but significant adjustment by writing

$$a = -|b|q + r = |b|(-1 - q) + (|b| - r).$$

Now $0 < r < |b|$ implies that $0 < |b| - r < |b|$ and so

$$a = bq' + r'$$

where $r' = |b| - r$ and $q' = -1 - q \ (b > 0)$, $q' = 1 + q \ (b < 0)$. This completes the proof. □

Examples 9

1. $a = -574$, $b = 34$. By the Division Algorithm (First Version) we have

$$574 = 34.16 + 30,$$

from which we have

$$-574 = 34.(-16) - 30 = 34(-16 - 1) + (34 - 30) = 34(-17) + 4$$

giving $q = -17$, $r = 4$.

2. $a = -6412$, $b = -97$. By the Division Algorithm (First Version) we have

$$6412 = 97.66 + 10,$$

from which we have

$$-6412 = (-97).66 - 10 = (-97).(66 + 1) + (97 - 10) = (-97).67 + 87$$

giving $q = 67$, $r = 87$.

Exercises 2.3

1. Given integers a and b, find integers q and r such that $a = bq + r$
   $(0 \leq r < |b|)$ where: (i) $a = 8$, $b = 6$; (ii) $a = 27$, $b = 8$; (iii) $a = 241$,
   $b = 35$; (iv) $a = -2513$, $b = 46$; (v) $a = 54321$, $b = -761$; (vi)
   $a = -52148$, $b = -732$.

2. Let x and y be real numbers. Prove (i) $|xy| = |x||y|$ and (ii)
   $|x + y| \leq |x| + |y|$. Construct an example in which, for appropriate
   choice of x and y, $|x - y| \geq |x| - |y|$.

2.4 Euclidean Algorithm

We now describe an effective method, the so-called Euclidean Algorithm, for
finding the G.C.D. of two integers. The Euclidean Algorithm depends on repeat-
ing the Division Algorithm for a finite number of times until a particular conclusion, yielding the G.C.D., is reached. The method appeared originally in a
geometrical form as Proposition 2 of Book VII of 'The Elements' – a book
which was composed by Euclid of Alexandria around 300 BC and which survived
as a textbook for over 2000 years.

First we prove a brief but useful result.

Theorem 3

Let a, b and d be non-zero integers and let d be a positive common divisor of a, b.
Suppose there exist integers x and y such that

$$ax + by = d.$$

Then d is the greatest common divisor of a and b.

Proof

Let c be a common divisor of a, b. We require to show that c divides d. Now we
certainly have $a = ca'$, $b = cb'$ for suitable a', $b' \in \mathbb{Z}$. Hence

$$d = ax + by = (ca')x + (cb')y = c(a'x) + c(b'y) = c(a'x + b'y).$$

Hence c divides d and so $d = (a, b)$.   □

Example 10

It is easy to verify that 97 is a common divisor of 291 and 388 but it is less easy, without the use of the Euclidean Algorithm, to find an equation of the form given in Theorem 3. One such equation is

$$291.15 + 388(-11) = 97,$$

and from this equation we may conclude that 97 is the G.C.D. of 191 and 388.

We shall give the explicit procedure of the Euclidean Algorithm but, by way of preliminary illustration, we begin with two particular examples of its use.

Examples 11

1. Suppose we wish to find the G.C.D. of 4947 and 1552. Our procedure is to apply the Division Algorithm systematically as follows. Applying the Division Algorithm three times we have

$$4947 = 1552.3 + 291 \tag{1}$$

$$1552 = 291.5 + 97 \tag{2}$$

$$291 = 97.3 \tag{3}$$

We claim that 97 is the required G.C.D. First we have to prove that 97 is indeed a divisor. We consider the equations in reverse order. Certainly from equation (3), 97 divides 291. But then from equation (2) 97 divides 1552. In turn, from equation (1), 97 divides 4947. Thus 97 is a common divisor of 1552 and 4947. Now to prove that 97 is the G.C.D. we let d be a divisor of 1552 and 4947. By equation (1) d divides 291 and then from (2) d divides 97. Hence 97 is the G.C.D.

We also write 97 in the form of Theorem 3 as follows. From equation (2)

$$97 = 1552 - 291.5$$

and hence, from equation (1),

$$97 = 1552 - (4947 - 1552.3).5 = 1552 - 4947.5 + 1552.3.5$$
$$= 4947.(-5) + 1552.16.$$

As soon as we have established that 97 is a common divisor then the representation of 97 in the form above gives an alternative proof that 97 is indeed the G.C.D.

2. Suppose we wish to find the G.C.D. of 163059 and 80004. Then applying the Division Algorithm four times we have

$$163059 = 80004.2 + 3051 \tag{1}$$

$$80004 = 3051.26 + 678 \tag{2}$$

$$3051 = 678.4 + 339 \tag{3}$$

$$678 = 339.2 \tag{4}$$

We consider these equations in the reverse order to show that 339 is a common divisor of 163059 and 80004.

From (4) 339 divides 3051. And so from (3) 339 divides 3051. But then (2) implies that 339 divides 80004 and then finally from (1), 339 divides 163059. Hence 339 is a common divisor of 163059 and 80004. By considering the equations in the normal order we may conclude, as in the previous Example, that 339 is, in fact, the G.C.D.

We may also establish this conclusion by writing 339 in the form $163059x + 80004y$ as follows. From (3) we have

$$339 = 3051 - 678.4$$

and so from (2) we have

$$339 = 3051 - (80004 - 3051.26).4 = 3051.105 - 80004.4$$

Applying (1) now gives

$$339 = (163059 - 80004.2)105 - 80004.4 = 163059.105 + 80004.(-214)$$

as we desired.

Keeping these two Examples in our minds we may now proceed to give the Euclidean Algorithm in its proper generality.

The Euclidean Algorithm

Given two non-zero integers a and b, we shall describe how the Algorithm will yield, in a finite number of steps, the G.C.D. of a and b. We follow the procedure of the two examples above. For convenience we may suppose $a > 0$, $b > 0$ and introduce the notation $a_0 = a$, $a_1 = b$.

By the Division Algorithm we obtain integers q_1 and a_2 such that

$$a_0 = a_1 q_1 + a_2, \quad 0 \le a_2 < a_1.$$

If $a_2 = 0$ then $a_0 = a_1 q_1$ and we have immediately that a_1 is the G.C.D. of a_0 and a_1. If $a_2 \neq 0$ then we continue as follows. We know, again by Division Algorithm, that we obtain integers q_2, a_3 such that

$$a_1 = a_2 q_2 + a_3, \quad 0 \leq a_3 < a_2.$$

If $a_3 = 0$ then we halt the procedure but if $a_3 \neq 0$ we continue and obtain integers q_3, a_4, such that

$$a_2 = a_3 q_3 + a_4, \quad 0 \leq a_4 < a_3.$$

Either $a_4 = 0$ or if $a_4 \neq 0$ then we continue as in the examples above.

$$a_0 = a_1 q_1 + a_2 \tag{1}$$

$$a_2 = a_2 q_2 + a_3 \tag{2}$$

$$a_2 = a_3 q_3 + a_4 \tag{3}$$

$$a_3 = a_4 q_4 + a_5 \tag{4}$$

Now continuing we obtain a finite or infinite sequence $a_1, a_2, a_3, \ldots$ of integers such that

$$a_1 > a_2 > a_3 > \ldots \geq 0.$$

This is only possible if the sequence is finite and for some n, $a_{n+1} = 0$. Thus suppose we halt at the nth equation. As in the examples above we have the following equations:

$$a_0 = a_1 q_1 + a_2 \tag{1}$$

$$a_1 = a_2 q_2 + a_3 \tag{2}$$

$$a_2 = a_3 q_3 + a_4 \tag{3}$$

$$a_3 = a_4 q_4 + a_5 \tag{4}$$

$$\cdots \quad \cdots \quad \cdots$$

$$a_{n-3} = a_{n-2} q_{n-2} + a_{n-1} \tag{$n-2$}$$

$$a_{n-2} = a_{n-1} q_{n-1} + a_n \tag{$n-1$}$$

$$a_{n-1} = a_n q_n \tag{n}$$

We claim that a_n is indeed the G.C.D. of $a_0 = a$ and $a_1 = b$. We prove first that a_n is a common divisor of a_0 and a_1 by considering the equations in the reverse order. From (n) a_n divides a_{n-1}. Then from $(n-1)$ a_n divides a_{n-2}. Continuing we have finally, on considering (1), that a_n divides a_2 and a_1 and so a_n divides a_0.

Thus a_n is a common divisor of a_0 and a_1. Let now c be a common divisor of a_0 and a_1. By (1) c divides $a_0 - a_1 q_1 = a_2$. Then from (2) c divides $a_1 - a_2 q_2 = a_3$. Continuing we find that c divides $a_{n-2} - a_{n-1} q_{n-1} = a_n$. Thus, as we claimed, a is the G.C.D. of a_0 and a_1.

We are also now able to write a_n in the form $a_0 x + a_1 y$ for some $x, y \in \mathbb{Z}$.

From $(n - 1)$

$$a_n = a_{n-2} - a_{n-1} q_{n-1}.$$

From $(n - 2)$

$$a_{n-1} = a_{n-3} - a_{n-2} q_{n-2}$$

and so, substituting from $(n - 2)$ into $(n - 1)$,

$$a_n = a_{n-2} - (a_{n-3} - a_{n-2} q_{n-2}) q_{n-1}$$
$$= (1 + q_{n-2} q_{n-1}) a_{n-2} - a_{n-3}.$$

From $(n - 3)$

$$a_{n-2} = a_{n-4} - a_{n-3} q_{n-3}$$

and so, substituting from $(n - 3)$ into the equation for a_n we have

$$a_n = (1 + q_{n-2} q_{n-1}) a_{n-2} - a_{n-3}$$
$$= (1 + q_{n-2} q_{n-1})[a_{n-4} - a_{n-3} q_{n-3}] - a_{n-3}$$
$$= [-q_{n-3}(1 + q_{n-2} q_{n-1}) - 1] a_{n-3} + (1 + q_{n-2} q_{n-1}) a_{n-4}$$
$$= -(1 + q_{n-3} + q_{n-3} q_{n-2} q_{n-1}) a_{n-3} + (1 + q_{n-2} q_{n-1}) a_{n-4}.$$

Continuing this procedure we shall eventually obtain an equation of the form

$$a_n = x a_0 + y a_1$$

for some $x, y \in \mathbb{Z}$.

We may summarize this last result in the form of a converse to Theorem 3.

Theorem 4

Let a and b be non-zero integers and let d be the greatest common divisor of a and b. Then there exist integers x and y such that

$$d = ax + by.$$

We remark in passing that x and y are not unique since $d = ax + by$ implies that $d = a(x - bt) + b(y + at)$ for any integer t.

We conclude this section with one further example which, in view of our extended discussion above, we shall give in as brief a form as is necessary for calculation.

Example 12

We wish to find the G.C.D. of 108810 and 93346 and to write the G.C.D. as $108810x + 93346y$ for some $x, y \in \mathbf{Z}$.

$$108810 = 93346.1 + 15464 \tag{1}$$

$$93346 = 15464.4 + 562 \tag{2}$$

$$15464 = 562.27 + 290 \tag{3}$$

$$562 = 290.1 + 272 \tag{4}$$

$$290 = 272.1 + 18 \tag{5}$$

$$272 = 18.15 + 2 \tag{6}$$

$$18 = 9.2 \tag{7}$$

Thus 2 is the G.C.D. of 108810 and 93346.

Reversing the order of the equations we have

$$
\begin{aligned}
(93346, 108810) = 2 &= 272 - 18.15 \text{ (from (6))} \\
&= 272 - (290 - 272).15 \text{ (from (5))} \\
&= -290.15 + 272.16 \\
&= -290.15 + (562 - 290).16 \text{ (from (4))} \\
&= 562.16 - 290.31 \\
&= 562.16 - (15464 - 562.27).31 \text{ (from (3))} \\
&= -15464.31 + 562.853 \\
&= -15464.31 + (93346 - 15464.6).853 \text{ (from (2))} \\
&= 93346.853 - 15464.5149 \\
&= 93346.853 - (108810 - 93346).5149 \text{ (from (1))} \\
&= 93346.6002 - 108810.5149
\end{aligned}
$$

which gives the desired result.

Exercises 2.4

1. Let a, b, x, y and n be integers such that n divides both a and b. Prove formally that n divides $ax + by$.

2. Use the Euclidean Algorithm to find the greatest common divisor (a, b) of each pair of given integers a and b and write (a, b) in the form $ax + by$ for suitable integers x and y: (i) $a = 32$, $b = 44$; (ii) $a = 150$, $b = 105$; (iii) $a = 3718$, $b = 1222$; (iv) $a = 96$, $b = 764$; (v) $a = 7224$, $b = 6214$; (vi) $a = 613640$, $b = 152881$.

3. Let a, b and n be integers. If n divides $a - b$ prove that n divides $a^2 - b^2$ and $a^3 - b^3$. If n divides $a + b$ prove that n divides $a^3 + b^3$ but does not necessarily divide $a^2 + b^2$.

2.5 Primes

We have defined a prime p as being an integer strictly greater than 1 which is divisible only by ± 1 or $\pm p$. It is now appropriate to examine some consequences of the division of an integer by a prime.

Definition 6

Let a and b be non-zero integers. Then a and b are said to be **coprime** (or to be prime to one another) if 1 is their greatest common divisor.

Example 13

$$60 \text{ and } 77 \text{ are coprime since } (60, 77) = 1.$$

It is immediate from the results above that if a and b are coprime then there exist $x, y \in \mathbf{Z}$ such that $ax + by = 1$. The converse is also true, namely that if for given $a, b \in \mathbf{Z}$, $a \neq 0$, $b \neq 0$, there exist $x, y \in \mathbf{Z}$ such that $ax + by = 1$ then $(a, b) = 1$.

We come now to a long-established and important result on division by a prime.

Theorem 5 ('The Elements', Book VII, Proposition 30)

Let a and b be integers and let p be a prime. Let p divide ab. Then either p divides a or p divides b.

Proof

For the sake of argument suppose p does not divide b. Then since the G.C.D. of p and b divides p the G.C.D. can only be 1. Hence there exist $x, y \in \mathbb{Z}$ such that $px + by = 1$. But then $apx + aby = a$. Clearly p divides apx and p divides ab (by hypothesis) and so p divides a. □

Applications of this result arise frequently. Using Theorem 5 we prove, as a matter of interest, the well-known result that $\sqrt{2} \notin \mathbb{Q}$, thus proving that the set of real numbers $\mathbb{R}$ is strictly greater than the set of rational numbers $\mathbb{Q}$.

Theorem 6

$$\sqrt{2} \notin \mathbb{Q}.$$

Proof

For the sake of argument suppose that $\sqrt{2} \in \mathbb{Q}$. Then we have $m, n \in \mathbb{N}$ such that $\sqrt{2} = \dfrac{m}{n}$. We may further suppose that m and n have no common divisor in $\mathbb{N}$ other than 1. Now $m^2 = 2n^2$ and so 2 divides m^2. From Theorem 5, 2 divides m and we may write $m = 2m'$ for some $m' \in \mathbb{N}$. Then $4(m')^2 = (2m')^2 = m^2 = 2n^2$ and so $2(m')^2 = n^2$. But this now implies that 2 divides n^2 and so n. We have arrived at a contradiction because we have deduced that 2 divides both m and n. Hence our initial supposition is wrong and so

$$\sqrt{2} \notin \mathbb{Q}.$$ □

Remark

The argument given above easily adapts to showing that $\sqrt{3}, \sqrt{5}, \ldots, \sqrt{p}$ (p prime) are not in $\mathbb{Q}$; indeed we may show that $\sqrt{n}$ ($n \in \mathbb{N}$), where n is not of the form m^2 ($m \in \mathbb{N}$), is not in $\mathbb{Q}$.

As reported in the Theaetetus, a dialogue of Plato (c. 429–347 BC), Theodorus (born c. 460 BC) was the first to prove that $\sqrt{3}, \sqrt{5}, \ldots, \sqrt{17}$ are irrational, the irrationality of $\sqrt{2}$ being apparently already known.

Example 14

Let a, $b \in \mathbf{Z}$ and let p be prime. Suppose p divides $a^2 + pb^2$. Then certainly p divides $(a^2 + pb^2) - pb^2$, and so p divides a^2 from which we conclude that p divides a.

In Theorem 5 it should be clear that the issue is not so much the primeness of p but the fact that p and b are coprime. Consequently, by essentially the same argument as in that theorem, we may obtain the following result.

Theorem 7

Let a and b be integers and let q be an integer which is prime to b. Let q divide ab. Then q divides a.

We give an application of this last result.

Example 15

Let a and b be integers having d as G.C.D. Then, as we recall from Theorem 4, there exist integers x and y such that $ax + by = d$.

We know that x and y are not unique. Suppose x' and y' are integers such that $ax' + by' = d$. Our natural curiosity should induce us to enquire whether there exists any relationship between the pair x and y and the pair x' and y'. Now we have $a = a'd$ and $b = b'd$ for some coprime integers a' and b'. But, as we have the equation $ax + by = d = ax' + by'$ it follows, on rearranging, that

$$a(x' - x) = -b(y' - y)$$

and, on dividing by d,

$$a'(x' - x) = -b'(y' - y)$$

But a' and b' are coprime and so, by Theorem 7, a' divides $y' - y$. Thus there exists an integer t such that

$$y' - y = a't.$$

But then we have

$$a'(x' - x) = -b'a't$$

and so

$$x' - x = -b't.$$

Thus, finally, the relationship we are seeking is:

$$x' = x - b't = x - \frac{b}{d}t,$$

$$y' = y + a't = y + \frac{a}{d}t,$$

where d is the G.C.D. of a and b.

It is important to be able to factorize integers into products of primes. For small numbers such as 1363 we can usually perform the factorization with a modest supply of pen and paper, for large numbers such as 45615161 a copious supply of pen and paper may enable a factorization to be found but for very large numbers such as $2^{64} + 1$ even an advanced calculator may be unable to perform the factorization. The reader is invited to factorize 1363 but may be relieved to be advised that

$$45615161 = 5879.7759, \text{ and } 2^{64} + 1 = 274177.67280421310721$$

when decomposed into prime factors.

A useful guide to the largeness of the prime that we need to consider in any attempted factorization of a possibly composite number is given by the next lemma.

Lemma 3

Let n be a composite integer. Then there is at least one prime p dividing n for which $p \leq \sqrt{n}$.

Proof

Let p be prime dividing n. If p^2 divides n then $p^2 \leq n$ and so $p \leq \sqrt{n}$. If p^2 does not divide n then there is a second prime q such that pq divides n. Changing notation, if necessary, we may suppose that $p < q$. Then $p^2 < pq \leq n$ and so $p < \sqrt{n}$. $\square$

Although possibly unable to perform an explicit factorization into prime factors we know nevertheless that one always exists and this fact is made precise in the next theorem, which on account of its importance, has been designated 'The Fundamental Theorem of Arithmetic'. Before formally stating the theorem we give an illustration.

Example 16

The number 5564832 is expressible as a product of primes in many ways, three of which are as follows:

$$5564832 = 2^2.7^2.13.3.7.2^3.13 = 7.13.2^2.7.3.7.2^3.13 = 2^5.3.7^3.13^2.$$

We assert that exactly the same primes 2, 3, 7, 13 will appear in any such factorization and that, furthermore, the powers 5, 1, 3, 2 to which respectively the primes 2, 3, 7, 13 appear, are determined solely by the number 5564832.

Theorem 8 The Fundamental Theorem of Arithmetic
('The Elements', Book IX, Proposition 14)

Every integer n, $n \in \mathbb{N}$ ($n \geq 2$) is expressible as a product of a finite number of primes. Furthermore if n is written as

$$n = p_1 p_2 \ldots p_r = q_1 q_2 \ldots q_s.$$

where $p_1, p_2, \ldots, p_r$; $q_1, q_2, \ldots, q_s$ are primes then $r = s$ and, with a suitable re-ordering if necessary, $p_i = q_i$ ($i = 1, 2, \ldots, r$).

Proof

If n is a prime and, in particular, if $n = 2$, then n is trivially a product of primes. Suppose $n > 2$. We shall argue by induction and suppose that the result is true for $n - 1$ and for all m such that $2 \leq m \leq n - 1$. We prove first that n is a product of primes.

If n is a prime then, as we remarked, the assertion is true. Suppose that n is composite and that $n = n_1 n_2$ where $n_1, n_2 \in \mathbb{N}$, $1 < n_1 < n$, $1 < n_2 < n$. By our assumption each of n_1 and n_2 is a product of primes and so therefore is n. Suppose now that

$$n = p_1 p_2 \ldots p_r = q_1 q_2 \ldots q_s.$$

Then p_1 divides $q_1 q_2 \ldots q_n$. By Theorem 5, p_1 must divide one of $q_1, q_2, \ldots, q_s$. By renumbering the q_i's if necessary we may suppose p_1 divides q_1. But p_1 and q_1 are primes so $p_1 = q_1$. Thus we have

$$p_1 p_2 \ldots q_r = p_1 q_2 q_3 \ldots q_s$$

from which

$$p_2 p_3 \ldots q_r = q_2 q_3 \ldots q_s$$

By our induction assumption we have $r = s$ and with suitable renumbering $p_i = q_i$ ($i = 2, 3, \ldots, n$). This completes the proof. $\square$

Arising out of this result we may collect together equal prime divisors and so we conclude that n may be written as

$$n = p_1^{\alpha_1} p_2^{\alpha_2} \ldots p_t^{\alpha_t}$$

where $p_1, p_2, \ldots, p_t$ are distinct primes and $\alpha_i > 0$ $(i = 1, 2, \ldots, t)$.

Many fascinating, but sometimes still unresolved, questions may be raised in regard to primes. Euclid answered one obvious initial question, namely 'how many?'

Theorem 9 ('The Elements', Book IX, Proposition 20)

There are infinitely many primes.

Proof

Let $p_1, p_2, \ldots$ be the primes in ascending order (of course $p_1 = 2$, $p_2 = 3$, $p_3 = 5$, etc.). We argue by contradiction and so suppose there is only a finite number N of primes; p_N is then the largest prime. Let

$$M = (p_1 p_2 \ldots p_N) + 1$$

M cannot be a prime since $M > p_N$. But, since M is a product of some of the primes $p_1, p_2, \ldots, p_N$, at least one of these primes must divide M, say p_1 divides M. But we have

$$\frac{M}{p_1} - p_2 p_3 \ldots p_N = \frac{1}{p_1}.$$

where the number on the left-hand side of this equation is an integer whereas the number on the right-hand side is not. We have reached a contradiction and so there is not a finite number of primes and we obtain the result. $\qquad \square$

Various simple extensions of this result are known. We shall give one such extension after some preliminary remarks.

Any odd integer is either of the form $4n + 1$ or $4n + 3$ for suitable $n \in \mathbb{Z}$, for example $21 = 4.5 + 1$ and $23 = 4.5 + 3$. In particular every prime other than 2 is of the form $4n + 1$ or $4n + 3$. We observe that the product of odd integers, all of which are of the form $4n + 1$, is again an integer of this form since

$$(4n_1 + 1)(4n_2 + 1) = 4(4n_1 n_2 + n_1 + n_2) + 1.$$

Example 17

The first ten primes, of the form $4n + 3$, are 3, 7, 11, 19, 23, 31, 43, 47, 59, 67.

Theorem 10

There are infinitely many primes of the form $4n + 3$.

Proof

Let $p_1, p_2, \ldots$ be the primes of the form $4n + 3$ in ascending order (of course $p_1 = 3$, $p_2 = 7$, $p_3 = 11$, etc.). We argue by contradiction and so suppose there is only a finite number N of such primes; p_N is then the largest such prime. Let

$$M = 4(p_2 p_3 \ldots p_N) + 3$$

Then, by an argument similar to that used in the proof of Theorem 9, M is not divisible by any of $p_2, p_3, \ldots, p_N$. Further M is not divisible by 3 since 3 does not divide $4 (p_2 p_3 \ldots p_N)$. Thus M is of the form $4n + 3$ but no prime factor of M is of this form. Thus all the prime factors of M must be of the form $4n + 1$. But from the remark above the product of such prime factors is again of the form $4n + 1$ and not of the form $4n + 3$. Thus we obtain a contradiction and so the result is proved.   □

We have developed some elementary properties of integers in order to whet the appetite for further study in number theory and, more importantly for our purposes, to prepare the ground for the axiomatic treatment of groups, rings and fields which is to follow. However, before leaving the present context we offer some random remarks on these objects of wonder the primes.

Sieve of Eratosthenes

Eratosthenes (c. 275–195 BC) gave a method for detecting primes which, for fairly obvious reasons, is known as a 'sieve'. The implementation of the method requires Lemma 3 above.

Lemma 3 implies that if we wish to determine whether a positive integer n is prime then we may confine our attention to possible prime divisors less than or equal to $\sqrt{n}$. We illustrate the method by detecting those primes less than 100. We write down the integers from 1 to 100 and successively eliminate those divisible by 2, 3, 5, 7, these being the primes $\leq \sqrt{100} = 10$ as follows. We circle 2 as being prime and stroke out every 2nd number thereafter. We circle 3 as the next prime and stroke out every 3rd number thereafter. Similarly we circle 5, and then 7, and stroke out every 5th, and then every 7th, number. We obtain the following picture:

1	②	③	4̶	⑤	6̶	⑦	8̶	9̶	1̶0̶
11	1̶2̶	13	1̶4̶	1̶5̶	1̶6̶	17	1̶8̶	19	2̶0̶
2̶1̶	2̶2̶	23	2̶4̶	2̶5̶	2̶6̶	2̶7̶	2̶8̶	29	3̶0̶
31	3̶2̶	3̶3̶	3̶4̶	3̶5̶	3̶6̶	37	3̶8̶	3̶9̶	4̶0̶
41	4̶2̶	43	4̶4̶	4̶5̶	4̶6̶	47	4̶8̶	4̶9̶	5̶0̶
5̶1̶	5̶2̶	53	5̶4̶	5̶5̶	5̶6̶	5̶7̶	5̶8̶	59	6̶0̶
61	6̶2̶	6̶3̶	6̶4̶	6̶5̶	6̶6̶	67	6̶8̶	6̶9̶	7̶0̶
71	7̶2̶	73	7̶4̶	7̶5̶	7̶6̶	7̶7̶	7̶8̶	79	8̶0̶
8̶1̶	8̶2̶	83	8̶4̶	8̶5̶	8̶6̶	8̶7̶	8̶8̶	89	9̶0̶
9̶1̶	9̶2̶	9̶3̶	9̶4̶	9̶5̶	9̶6̶	97	9̶8̶	9̶9̶	1̶0̶0̶

The numbers without strokes, other than 1, are 11, 13, 17, 19, 23, 29, 31, 37, 41, 43, 47, 53, 59, 61, 67, 71, 73, 79, 83, 89, 97 which are therefore together with 2, 3, 5, 7 the primes less than 100.

The method, while systematic, has obvious limitations.

The Distribution of the Primes

The primes are distributed irregularly amongst the integers but, nevertheless, we should like some measure of the irregularity. Two results which are of especial interest will be quoted without proofs (which are analytic rather than algebraic).

Let $p_1, p_2, \ldots$ be the ascending sequence of primes. J. Bertrand surmised (1845) and P.L. Chebyshev (1821–94) proved that

$$p_{n+1} < 2p_n \quad (n \in \mathbb{N}).$$

For the second result we introduce a function commonly denoted by π (thus π is here not 3.14159...) and defined for $x \in \mathbb{N}$ by letting $\pi(x)$ be the number of primes $\leq x$. Quick calculation shows that $\pi(1) = 0$, $\pi(2) = 1$, $\pi(3) = \pi(4) = 2$, $\pi(5) = \pi(6) = 3$, $\pi(7) = \pi(8) = \pi(9) = \pi(10) = 4$, etc. K.F. Gauss (1777–1855) formulated the remarkable conjecture that

$$\frac{\pi(x) \log x}{x} \to 1 \text{ as } x \to \infty.$$

This was established independently by C.J. de la Valle-Poussin (1866–1962) and J. Hadamard (1865–1963), the longevity of whom cannot solely be attributed to an interest in number theory.

Goldbach's Conjecture

This conjecture originated with C. Goldbach (1690–1764) and is to the effect that every even integer, other than 2, is the sum of two primes; for example, $4 = 2 + 2$, $6 = 3 + 3$, $8 = 5 + 3$, $10 = 3 + 7 = 5 + 5$. The difficulty of tackling the conjecture seems to lie in the fact that primes are clearly concerned with factorization but not obviously with addition. Various allied results have been obtained but the conjecture itself remains to tantalize.

After this brief excursion into aspects of number theory we shall develop the previously mentioned axiomatic treatment through a discussion of polynomials.

Exercises 2.5

1. Write down formal proofs that $\sqrt{3}$ and $\sqrt{6}$ are irrational.

2. Prove that $\sqrt{2} + \sqrt{3}$ is irrational.

3. The sum of two rational numbers is rational. Is the sum of two irrational numbers always irrational?

4. Write down a formal proof of Theorem 7.

5. Two primes p and q are said to be a 'prime pair' if $q = p + 2$, for example 17 and 19 is a prime pair. How many prime pairs are there between 1 and 100? (It is unknown whether or not there are infinitely many prime pairs.)

6. Apply the Sieve of Eratosthenes to find the primes between 100 and 200.

7. Calculate $\pi(10n), n = 1, 2, \ldots, 10$.

8. In how many ways may 144 be written as the sum of two primes?

3
Introduction to Rings

In the previous chapter we have seen that the integers possess a division algorithm and that from this division algorithm there may be derived a Euclidean Algorithm for finding the greatest common divisor of two given integers. 'Polynomials' share many properties in common with the integers, having a division algorithm and a corresponding Euclidean Algorithm. As our treatment of polynomials proceeds, initially somewhat informally, it will become apparent that we need to consider much more precisely the extent to which integers and polynomials share common features. In this way we shall be led to enunciate axioms for an algebraic system called a 'ring' and for a ring of a particular type called an 'integral domain' which incorporates some of the features common to integers and polynomials.

Axioms in mathematics are never constructed arbitrarily but are designed to focus attention on significant aspects of the system or systems under consideration. While axioms are famously present in Euclid's 'Elements' their modern extensive use stems from the mathematical work of the nineteenth century, a decisive influence in their use being that of the great D. Hilbert (1862–1943). The concept of a ring was introduced by R. Dedekind (1831–1916) but the first set of axioms for a ring, although not quite equivalent to those in use today, was published in 1914 by A.H. Fraenkel (1891–1965).

3.1 Concept of a Polynomial

What does the term 'polynomial' mean to us? We think of algebraic expressions such as $2x + 1$ or $x^2 + 5x + 6$ or $2x^3 + x^2 - 2x - 1$ etc. We speak of these as being 'polynomials in x' and say that $2x + 1$ has 'degree' 1, $x^2 + 5x + 6$ has 'degree' 2 and $2x^3 + x^2 - 2x - 1$ has 'degree' 3. We know how to add, subtract and multiply such polynomials:

$$(3x + 5) + (4x^2 - 2x - 1) = 4x^2 + x + 4,$$

$$(2x^2 - 6) - (x^2 - 5x - 1) = x^2 + 5x - 5,$$

$$(x^2 + 2x + 4)(3x + 1) = x^2(3x + 1) + 2x(3x + 1) + 4(3x + 1)$$

$$= (3x^3 + x^2) + (6x^2 + 2x) + (12x + 4)$$

$$= 3x^3 + 7x^2 + 14x + 4.$$

As in the case of the integers we may perform long division, for example $3x + 1$ does not divide $9x^3 - 3x^2 + 6x + 4$ but leaves a remainder when we employ long division as follows:

$$
\begin{array}{r}
3x^2 - 2x + \frac{8}{3} \\
3x + 1 \overline{)\,9x^3 - 3x^2 + 6x + 4} \\
9x^3 + 3x^2 \\
\hline
-6x^2 + 6x \\
-6x^2 - 2x \\
\hline
8x + 4 \\
8x + \frac{8}{3} \\
\hline
\frac{4}{3}
\end{array}
$$

Thus we write

$$\frac{9x^3 - 3x^2 + 6x + 4}{3x + 1} = 3x^2 - 2x + \frac{8}{3} + \frac{\frac{4}{3}}{3x + 1},$$

or, more usefully,

$$9x^3 - 3x^2 + 6x + 4 = (3x + 1)\left(3x^2 - 2x + \frac{8}{3}\right) + \frac{4}{3},$$

which is in a form to be expected from the existence of a division algorithm. For convenience we gather together in a more formal manner some of the aspects of polynomials mentioned above.

Definition 1

By a **polynomial in x over** $\mathbb{Q}$ we understand an expression $a(x)$ of the form

$$a(x) = a_0 + a_1 x + \ldots + a_m x^m$$

where the so-called **coefficients** $a_0, a_1, \ldots, a_m$ are rational numbers, and x is sometimes called an **'indeterminate'**. If all of the coefficients are 0 the polynomial is said to be the **zero polynomial**, denoted by 0. If $a(x)$ is not the zero polynomial and if $a_m \neq 0$ then $a(x)$ is said to have **degree** m, briefly $\deg a(x) = m$. If $a(x)$ has degree m and $a_m = 1$ then $a(x)$ is said to be **monic**. A polynomial which is either the zero polynomial or has degree 0 is said to be a **constant polynomial** and is therefore of the form $a(x) = a_0$ ($a_0 \in \mathbb{Q}$). Polynomials of degrees 1, 2 or 3 are sometimes called **linear**, **quadratic** or **cubic** respectively.

Two polynomials $a(x)$ and $b(x)$ where

$$a(x) = a_0 + a_1 x + \ldots + a_m x^m,$$
$$b(x) = b_0 + b_1 x + \ldots + b_n x^n,$$

are deemed to be **equal** if and only if $m = n$ and $a_0 = b_0, a_1 = b_1, \ldots, a_n = b_n$. In particular $a(x) = 0$ if and only if all coefficients of $a(x)$ are 0. The addition and multiplication of $a(x)$ and $b(x)$ are defined as

$$a(x) + b(x) = (a_0 + b_0) + (a_1 + b_1)x + \ldots,$$

$$a(x)b(x) = \left(\sum_{r=0}^{m} a_r x^r \right) \left(\sum_{s=0}^{n} b_s a^s \right)$$

$$= \sum_{r=0}^{m} \sum_{s=0}^{n} a_r b_s x^{r+s}$$

$$= a_0 b_0 + (a_0 b_1 + a_1 b_0)x + (a_0 b_2 + a_1 b_1 + a_2 b_0)x^2$$
$$+ \ldots + a_m b_n x^{m+n}.$$

The set of all polynomials in x over $\mathbb{Q}$ is denoted by $\mathbb{Q}[x]$. Similarly $\mathbb{Z}[x]$, $\mathbb{R}[x]$, $\mathbb{C}[x]$ are defined by choosing the relevant coefficients for the polynomials.

Example 1

$1 + \sqrt{2}x^2 + \sqrt{3}x^5$ is a polynomial in x over $\mathbb{R}$ of degree 5. $5 + 2y + 3y^2$ is a quadratic polynomial in y over $\mathbb{Z}$ (and also over $\mathbb{Q}$, $\mathbb{R}$ and $\mathbb{C}$). $2 + \frac{1}{2}z^2 + z^3$ is a monic cubic polynomial in z over $\mathbb{Q}$.

A key property of polynomials in x over $\mathbb{Q}$, say, is the possibility of substituting for the x. We may replace x in any equation involving polynomials in x by any number, not necessarily in $\mathbb{Q}$, and thereby obtain a valid equation involving that number. Thus if $a(x) = a_0 + a_1x + \ldots + a_mx^m \in \mathbb{Q}[x]$ and if $c \in \mathbb{R}$ we have $a(c) = a_0 + a_1c + \ldots + a_mc^m \in \mathbb{R}$ and if $f(x)$, $g(x)$, $h(x) \in \mathbb{Q}[x]$ are such that $f(x)g(x) = h(x)$ then $f(c)g(c) = h(c)$.

Example 2

Let $f(x) = x^2 + x + 1$, $g(x) = 2x - 3$, $h(x) = 2x^3 - x^2 - x - 3$. Then we may verify that $f(x)g(x) = h(x)$. If we substitute $\sqrt{2}$ for x then

$$f(\sqrt{2}) = (\sqrt{2})^2 + \sqrt{2} + 1 = 3 + \sqrt{2},$$

$$g(\sqrt{2}) = 2\sqrt{2} - 3, \quad h(\sqrt{2}) = 2(\sqrt{2})^3 - (\sqrt{2})^2 - \sqrt{2} - 3$$

$$= 3\sqrt{2} - 5,$$

and, as expected,

$$f(\sqrt{2})g(\sqrt{2}) = h(\sqrt{2}).$$

We shall investigate polynomials in $\mathbb{Q}[x]$. With obvious and appropriate changes our results apply also to $\mathbb{R}[x]$ and $\mathbb{C}[x]$ but not always to $\mathbb{Z}[x]$.

Definition 2

Let $f(x)$ and $g(x)$ be polynomials in $\mathbb{Q}[x]$, $g(x) \neq 0$. Then we say that $g(x)$ **divides** $f(x)$, or that $g(x)$ is a **divisor** of $f(x)$, if there exists a polynomial $h(x) \in \mathbb{Q}[x]$ such that $f(x) = g(x)h(x)$.

Notice that the zero polynomial 0 is divisible by any polynomial $g(x) \neq 0$ since $0 = g(x)0$ and that any polynomial $f(x) \neq 0$ has the **trivial divisors** c and $cf(x)$ for any $c \in \mathbb{Q}$, $c \neq 0$.

Examples 3

1. $2x^2 + 2 \in \mathbb{Q}[x]$ has the divisors c and $c(x^2 + 1)$ $(c \neq 0)$ since

$$2x^2 + 2 = \frac{2}{c}c(x^2 + 1).$$

2. $2x^2 + 3x + 1$ divides $2x^3 + x^2 - 2x - 1$ since

$$2x^3 + x^2 - 2x - 1 = (2x^2 + 3x + 1)(x - 1).$$

3. In $\mathbb{Q}[x]$, $x^2 - 2$ has only the trivial divisors c, $c(x^2 - 2)$, $c \in \mathbb{Q}$, $c \neq 0$. In $\mathbb{R}[x]$ however there is the non-trivial factorization

$$x^2 - 2 = (x - \sqrt{2})(x + \sqrt{2}).$$

This example shows that it is important to know in which system the factorization is to take place.

Lemma 1

Let $f(x)$, $g(x)$ and $h(x)$ be polynomials in $\mathbb{Q}[x]$ such that $g(x)$ and $h(x)$ are non-zero. Let $g(x)$ divide $f(x)$ and $h(x)$ divide $g(x)$. Then $h(x)$ divides $f(x)$.

Proof

This follows closely the proof of the corresponding lemma in Chapter 2.

Definition 3

Let $f(x)$ and $g(x)$ be non-zero polynomials in $\mathbb{Q}[x]$. A polynomial $h(x)$ is a **common divisor** of $f(x)$ and $g(x)$ if $h(x)$ divides $f(x)$ and $h(x)$ divides $g(x)$. A common divisor $d(x)$ which is monic and which is such that $d(x)$ is divisible by any common divisor of $f(x)$ and $g(x)$ is called the **greatest common divisor** (G.C.D.) of $f(x)$ and $g(x)$.

Example 4

The quadratic polynomials $2x^2 + 7x + 3$ and $6x^2 + x - 1$ in $\mathbb{Q}[x]$ have G.C.D. $x + \frac{1}{2}$ since

$$2x^2 + 7x + 3 = (2x + 1)(x + 3) = 2\left(x + \frac{1}{2}\right)(x + 3),$$

$$6x^2 + x - 1 = (2x + 1)(3x - 1) = 2\left(x + \frac{1}{2}\right)(3x - 1).$$

Exercises 3.1

In each of the following find the G.C.D. of the pair of polynomials

1. $x^2 + x - 2$, $x^2 + 3x + 2$.

2. $6x^2 + x - 2$, $15x^2 + 13x + 2$.

3. $x^3 + x^2 - x - 1, \quad x^3 + 2x^2 + x.$

4. $x^3 + x^2 + x + 1, \quad 2x^3 - x^2 + 2x - 1.$

3.2 Division and Euclidean Algorithms

As we remarked above, we shall prove results only for $\mathbb{Q}[x]$ but they apply with trivial modifications for $\mathbb{R}[x]$ and $\mathbb{C}[x]$.

Theorem 1 The Division Algorithm

Let $a(x)$ and $b(x)$ be polynomials over $\mathbb{Q}[x]$, $b(x)$ being non-zero. Then there exist polynomials $q(x)$ and $r(x)$ with coefficients in $\mathbb{Q}$ satisfying the condition

$$a(x) = b(x)q(x) + r(x)$$

where either $r(x) = 0$ or if $r(x) \neq 0$ then the degree of $r(x)$ is strictly less than the degree of $b(x)$.

Proof

Let $a(x)$ and $b(x)$ be the polynomials of degree m and n respectively given by

$$a(x) = a_0 + a_1 x + \ldots + a_m x^m, \quad a_m \neq 0,$$
$$b(x) = b_0 + b_1 x + \ldots + b_n x^n, \quad b_n \neq 0.$$

We first dispose of a trivial case, namely deg $a(x) <$ deg $b(x)$. We simply write

$$a(x) = 0b(x) + r(x)$$

where $q(x) = 0$ and $r(x) = a(x)$, which gives the result.

We shall use the Principle of Induction by arguing in regard to deg $a(x)$.

If deg $a(x) = 0$ then $\mathrm{a}(x) = a_o$. If also deg $a(x) =$ deg $b(x)$ then $b(x) = b_0$ and so

$$a(x) = b(x)q$$

where $q = \dfrac{a_0}{b_0}$. On the other hand if deg $a(x) <$ deg $b(x)$ then we have the result by the trivial case above.

Let now deg $a(x) > 0$. Suppose the result is true for all polynomials of degrees strictly less than the degree of $a(x)$. We wish to prove the result for $a(x)$. If deg $a(x) <$ deg $b(x)$ then again we have the trivial case above. Suppose therefore that deg $b(x) \leq$ deg $a(x)$. Then we aim to construct a polynomial of degree strictly less than deg $a(x)$ and to apply the induction assumption to this polynomial. Let

$$f(x) = a(x) - \frac{a_m}{b_n} x^{m-n} b(x) \quad \text{(by convention } x^0 = 1\text{).}$$

Then $f(x)$ has degree strictly less than deg $a(x)$ since the terms in x^m cancel as we see below:

$$f(x) = (a_0 + a_1 x + \ldots a_m x^m) - \frac{a_m}{b_n} (b_0 + b_1 x + \ldots + b_n x^n) x^{m-n}.$$

By the induction assumption there exist polynomials $p(x)$ and $r(x)$ such that

$$f(x) = b(x)p(x) + r(x)$$

where either $r(x) = 0$ or if $r(x) \neq 0$ then deg $r(x) <$ deg $b(x)$. Hence

$$b(x)p(x) + r(x) = f(x) = a(x) - \frac{a_m}{b_n} x^{m-n} b(x)$$

and so

$$a(x) = b(x)\left(p(x) + \frac{a_m}{b_n} x^{m-n} \right) + r(x)$$

$$= b(x)q(x) + r(x)$$

where $q(x) = p(x) + \frac{a_m}{b_n} x^{m-n}$ and either $r(x) = 0$ or if $r(x) \neq 0$ then deg $r(x) <$ deg $b(x)$. This completes the proof. □

The following useful result comes as a consequence of the Division Algorithm.

Theorem 2 The Remainder Theorem

Let $f(x) \in \mathbb{Q}[x]$ and let $c \in \mathbb{Q}$.

1. There exists $q(x) \in \mathbb{Q}[x]$ such that $f(x) = (x - c)q(x) + f(c)$.
2. $x - c$ divides $f(x)$ if and only if $f(c) = 0$.

Proof

1. By the Division Algorithm there exist $q(x)$ and $r(x)$ such that

$$f(x) = (x - c)q(x) + r(x)$$

where either $r(x) = 0$ or if $r(x) \neq 0$ then deg $r(x) <$ deg $(x - c) = 1$. In either case $r(x) = r_0$ where r_0 is a, possibly zero, constant. Then

$$f(c) = (c - c)q(c) + r_0 = 0q(c) + r_0 = r_0.$$

2. 2 is an immediate consequence of 1. □

Example 5

$$f(x) = 2x^5 + x^4 + 7x^3 + 2x + 10$$

is divisible by $x + 1$ since

$$f(-1) = 2(-1)^5 + (-1)^4 + 7(-1)^3 + 2(-1) + 10$$

$$= -2 + 1 - 7 - 2 + 10 = 0.$$

Examples 6

For the polynomials $a(x)$ and $b(x)$ below we find $q(x)$ and $r(x)$ such that

$$a(x) = b(x)q(x) + r(x)$$

where $r(x) = 0$ or if $r(x) \neq 0$ then deg $r(x) <$ deg $b(x)$. We employ long division which is, in fact, the basis of the Division Algorithm.

1. $a(x) = x^3 + 4x^2 + 5x + 7, b(x) = x + 1$.

$$
\require{enclose}
\begin{array}{r}
x^2 + 3x + 2 \\
x + 1 \enclose{longdiv}{x^3 + 4x^2 + 5x + 7} \\
\underline{x^3 + x^2} \\
3x^2 + 5x \\
\underline{3x^2 + 3x} \\
2x + 7 \\
\underline{2x + 2} \\
5
\end{array}
$$

Thus $a(x) = b(x)q(x) + r(x)$ where $q(x) = x^2 + 3x + 2$, $r(x) = 5$.

2. $a(x) = 2x^3 + 7x^2 + 1$, $b(x) = 3x + 2$

$$\begin{array}{r}
\frac{2}{3}x^2 + \frac{17}{9}x - \frac{34}{27} \\
3x + 2 \overline{)2x^3 + 7x^2 \qquad\qquad + 1} \\
2x^3 + \frac{4}{3}x^2 \\
\hline
\frac{17}{3}x^2 \\
\frac{17}{3}x^2 + \frac{34}{9}x \\
\hline
\frac{34}{9}x + 1 \\
\frac{34}{9}x - \frac{68}{27} \\
\hline
\frac{95}{27}
\end{array}$$

Thus

$$a(x) = b(x)q(x) + r(x)$$

where $q(x) = \frac{2}{3}x^2 + \frac{17}{9}x - \frac{34}{27}$, $r(x) = \frac{95}{27}$.

We turn now to the matter of finding the G.C.D. of two polynomials. The Euclidean Algorithm used here to find the G.C.D. is exactly analogous to the Euclidean Algorithm used in the case of the integers, we merely have to modify some details appropriately. In these circumstances we shall content our-selves therefore by illustrating the method by means of an example.

Examples 7

1. Suppose we wish to find the G.C.D. of

$$a(x) = 2x^3 - 5x^2 - 2x - 3 \text{ and } b(x) = x^3 - x^2 - x - 15.$$

Then as in the Euclidean Algorithm for finding the G.C.D. of two integers we use the Division Algorithm which in this example must be applied three times as follows:

$$2x^3 - 5x - 2x - 3 = 2(x^3 - x^2 - x - 15) + (-3x^2 + 27),$$

$$x^3 - x^2 - x - 15 = (-3x^2 + 27)\left(-\frac{1}{3}x + \frac{1}{3}\right) + (8x - 24),$$

$$-3x^2 + 27 = (8x - 24)\left(-\frac{1}{8}x - \frac{9}{8}\right).$$

Thus we would expect that $8x - 24$ would be 'almost' the G.C.D. However, it will be recalled that, for definiteness, we have chosen that the G.C.D. should be monic. But $8x - 24 = 8(x - 3)$ and so $x - 3$ is the required G.C.D. Furthermore by considering the above equations in reverse order we have

$$8(x - 3) = (x^3 - x^2 - x - 15) - (-3x^2 + 27)\left(-\frac{x}{3} + \frac{1}{3}\right)$$

$$= b(x) - [a(x) - 2b(x)]\left(-\frac{x}{3} + \frac{1}{3}\right)$$

$$= a(x)\left[-\left(-\frac{x}{3} + \frac{1}{3}\right)\right] + b(x)\left[1 + 2\left(-\frac{x}{3} + \frac{1}{3}\right)\right]$$

$$= a(x)\left[\frac{x}{3} - \frac{1}{3}\right] + b(x)\left[-\frac{2x}{3} + \frac{5}{3}\right],$$

from which we have

$$x - 3 = a(x)\frac{1}{24}(x - 1) + b(x)\frac{1}{24}(-2x + 5).$$

2. Suppose we are given the polynomials $a(x) = 2x^4 + 3x^3 + 5x^2 + 6x + 2$ and $b(x) = x^4 + 5x^2 + 6$, and we wish to find the G.C.D. $d(x)$ and to write $d(x)$ in the form

$$d(x) = a(x)f(x) + b(x)g(x)$$

for suitable polynomials $f(x)$ and $g(x)$.

We have

$$a(x) = 2b(x) + (3x^3 - 5x^2 + 6x - 10),$$

$$b(x) = (3x^3 - 5x^2 + 6x - 10)\left(\frac{x}{3} + \frac{5}{9}\right) + \left(\frac{52}{9}x^2 + \frac{104}{9}\right),$$

$$3x^3 - 5x^2 + 6x - 10 = \left(\frac{52}{9}x^2 + \frac{104}{9}\right)\left(\frac{27}{52}x - \frac{45}{52}\right).$$

Then, since

$$\frac{52}{9}x^2 + \frac{104}{9} = \frac{52}{9}(x^2 + 2),$$

the G.C.D. is $x^2 + 2$. Also

$$x^2 + 2 = \frac{9}{52}\left(\frac{52}{9}x^2 + \frac{104}{9}\right)$$

$$= \frac{9}{52}\left[b(x) - (3x^3 - 5x^2 + 6x - 10)\left(\frac{x}{3} + \frac{5}{9}\right)\right]$$

$$= \frac{1}{52}[9b(x) - (3x^3 - 5x^2 + 6x - 10)(3x + 5)]$$

$$= \frac{1}{52}[9b(x) - (a(x) - 2b(x))(3x + 5)]$$

$$= \frac{-(3x + 5)}{52}a(x) + \frac{(6x + 19)}{52}b(x).$$

Thus

$$f(x) = \frac{-(3x + 5)}{52}, \quad g(x) = \frac{6x + 19}{52}$$

is a possible solution for $f(x)$ and $g(x)$. (Recall that we know that $f(x)$ and $g(x)$ are not uniquely determined.)

Exercises 3.2

1. For the given polynomials $a(x)$ and $b(x)$ in $\mathbb{Q}[x]$, find $q(x)$ and $r(x)$ in $\mathbb{Q}[x]$ such that

$$a(x) = b(x)q(x) + r(x)$$

where either $r(x) = 0$ or if $r(x) \neq 0$ then deg $r(x) <$ deg $b(x)$.

(i) $a(x) = x^3 + 3x^2 + 1$, $b(x) = x^4 + 1$.
(ii) $a(x) = 3x^3 + 3x^2 + 2x + 1$, $b(x) = 3x^2 + 2$.
(iii) $a(x) = 2x^2 + 5x + 1$, $b(x) = 5x - 1$.
(iv) $a(x) = x^3 + 4x^2 + 4x + 1$, $b(x) = x + 3$.
(v) $a(x) = x^4 + 3x^2 + 1$, $b(x) = 2x^2 + 1$.

2. For the given polynomials $a(x)$ and $b(x)$ in $\mathbb{Q}[x]$, find the G.C.D. $d(x)$ and write $d(x)$ in the form

$$d(x) = a(x)f(x) + b(x)g(x)$$

for suitable polynomials $f(x)$ and $g(x)$ in $\mathbb{Q}[x]$.

(i) $a(x) = x^4 + x^3 + x + 1$, $b(x) = x^2 + x + 1$.
(ii) $a(x) = 2x^3 + 10x^2 + 2x + 10$, $b(x) = x^3 - 2x^2 + x - 2$.
(iii) $a(x) = x^4 - 4x^3 + 2x - 4$, $b(x) = x^3 + 2$.
(iv) $a(x) = x^3 + 5x^2 + 7x + 2$, $b(x) = x^3 + 2x^2 - 2x - 1$.

3.3 Axioms and Rings

From our deliberations we have seen that the system of polynomials behaves in a similar manner to the system of the integers; it would therefore seem advisable to study further whatever aspects these two systems have in common. In order to understand better what apparently similar systems have in common it is often convenient to lay down a common set of axioms which is satisfied by the systems. A set of axioms is not invented arbitrarily but is constructed to reproduce aspects of common and significant interest across the systems. Once the axioms are laid down the axiomatic system then delineated takes on, as it were, a life of its own and may then be explored both for its own sake and to achieve a better overall understanding of previously considered systems.

In regard to the integers and polynomials we observe that there are in each system two crucial operations, one is the **addition** of two numbers or polynomials and the other is the **multiplication** of two numbers or polynomials. These operations are said to be **binary** since they involve two numbers or polynomials. The operations satisfy certain conditions which are tacitly assumed in elementary treatments but which must be made explicit when we pass to an axiomatic treatment. We shall give appropriate axioms for the system, later defined to be a **ring**, in which both of these operations occur. For convenience of reference and in committing the axioms to memory, we give the definition of a ring in a succinct form but then follow the definition with a commentary upon the axioms of the definition. Finally, we conclude with the definition of an **integral domain** which is a ring of a particular kind and which brings together many of the more obvious features of integers and polynomials.

Definition 4

Let R be a non-empty set in which there are defined two binary operations called **addition** and **multiplication**. For a, $b \in R$ the outcome of the addition of a and b, called the **sum** of a and b, is denoted by $a + b$ ('a plus b') and the outcome of the multiplication of a by b, called the **product** of a by b, is denoted by the simple juxtaposition ab ('a times b'). Then R is called a **ring** if the following axioms hold.

1. Axioms of Addition

    1.1   For all $a, b \in R$, $a + b \in R$ (closure).

    1.2   For all $a, b, c \in R, (a + b) + c = a + (b + c)$ (associativity).

    1.3   There exists a distinguished element denoted by 0 such that for all $a \in R$, $a + 0 = 0 + a = a$ (0 is 'zero').

1.4 For any $a \in R$ there exists an element denoted by $-a \in R$ such that $a + (-a) = (-a) + a = 0$ (the inverse of addition).

1.5 For all $a, b \in R, a + b = b + a$ (commutativity).

2. Axioms of Multiplication

2.1 For all $a, b \in R$, $ab \in R$ (closure).

2.2 For all $a, b, c \in R$, $(ab)c = a(bc)$ (associativity).

3. Axioms of Distributivity

3.1 For all $a, b, c \in R, a(b + c) = ab + ac$ (distributivity).

3.2 For all $a, b, c \in R, (a + b)c = ac + bc$ (distributivity).

Commentary

1.1 The closure of addition is to ensure that the sum of a and b also belongs to R. It would be a bizarre system if this fact were not to be so.

1.2 The associativity of addition enables us to dispense with brackets in addition. Thus we may write simply $a + b + c$ where the meaning is $(a + b) + c$ or, equally, $a + (b + c)$.

1.3 This axiom only states that a zero exists. On the face of it there could be more than one such zero. However, if $0'$ were to denote a second zero we would have for all $a \in R$,

$$a + 0 = 0 + a = a, \quad a + 0' = 0' + a = a.$$

In particular we would have

$$0' + 0 = 0 + 0' = 0', \quad 0 + 0' = 0' + 0 = 0$$

from which $0' = 0$. Henceforth we may speak of **the zero** of R.

1.4 Now that we know that the zero element is unique, by carefully applying the axioms we may establish that for each $a \in R$ the inverse of addition $-a$ is uniquely determined by a; if we suppose that we have $b \in R$ such that $a + b = b + a = 0$ then we may show that $b = -a$ as follows.

By 1.2 (associativity)

$$b + (a + (-a)) = (b + a) + (-a).$$

But

$$b + (a + (-a)) = b + 0 = b$$

and

$$(b + a) + (-a) = 0 + (-a) = (-a),$$

from which we have $b = -a$. Henceforth we may speak of **the inverse of addition** $-a$ for given $a \in R$. Notice that as $a + (-a) = (-a) + a = 0$ we have $a = -(-a)$.

1.5   This axiom states that the sum of a and b is independent of the order in which a and b are added to one another; a and b are said to 'commute'. (The words 'commutativity' and 'commute' stem from electrical engineering in which a commutator is an apparatus for reversing the current.)

2.1   The closure of multiplication is to ensure that the product of a and b belongs to R (see comment on 1.1). Note, however, that we must maintain the order of a followed by b since, in general, ab and ba are not necessarily equal.

2.2   The associativity of multiplication enables us to dispense with brackets in multiplication. Thus abc is defined unambiguously, being either $(ab)c$ or $a(bc)$.

3.1/3.2 These axioms give conditions linking addition and multiplication. (The reader may recall the remarks in Chapter 2 on Goldbach's Conjecture.)

In these axioms there are certain tacit assumptions in regard to the order in which the bracketing and the two binary operations are to be considered. Thus the axiom $a(b + c) = ab + ac$ means that, on the left-hand side, we add b and c within the brackets to give $b + c$ and then we perform a multiplication by a whereas, on the right-hand side, we multiply a and b and also multiply a and c before performing the addition of ab and ac. These remarks may seem superfluous but only because we are so accustomed to the particular order of precedence of brackets and operations (an order which in years gone by was encapsulated in the mnemonic BODMAS = brackets/of/division/multiplication/ addition/subtraction). For example, by $a + bc$, we understand that b and c are first multiplied to give bc and then a and bc are added.

A convention also arises in regard to the minus sign '$-$'. In writing $a - b$ we are actually simply writing a followed by the additive inverse of b; however by $a - b$ we understand that in fact we intend $a + (-b)$, the $+$ sign being implied but normally omitted.

If we now relate these axioms either to the set of integers $\mathbb{Z}$ or to the set of polynomials $\mathbb{Q}[x]$ with the usual addition and multiplication operations, we should observe that the axioms are clearly satisfied; for example, the integers satisfy 1.5 since we have assumed from early childhood that the sum of two integers was the same irrespective of the order in which they were added. Nevertheless the reader should convince himself or herself, certainly for the integers and possibly for polynomials, that the other axioms are equally obviously satisfied. We shall henceforth speak of the 'ring of integers' and of the 'ring of polynomials' (although this latter terminology is still a trifle imprecise).

Example 8

Rings arise in many ways. Below we consider a ring the elements of which are mappings. Let X be a non-empty set. Let R be the set of mappings from X to $\mathbb{R}$. We introduce an addition and multiplication, denoted by '.', into R as follows. Let $f : X \to \mathbb{R}$ and $g : X \to \mathbb{R}$. Define the sum $f + g$ and the product $f.g$ as mappings from X to $\mathbb{R}$ by,

$$(f + g)(x) = f(x) + g(x) \quad (x \in X),$$

$$(f.g)(x) = f(x)g(x) \quad (x \in X).$$

We note that on the right-hand side of the first equation the plus sign $+$ refers to addition in $\mathbb{R}$ and on the right-hand side of the second equation the product is in $\mathbb{R}$. Consequently the sum and product of the mappings f and g are well-defined. We claim that R is a ring with the given definitions of sum and product. We require to verify all the axioms for a ring. We use the same numbering system as above.

1.1 Certainly, by definition, R is closed under addition.

1.2 Let f, g, $h \in R$. We want to show that $(f + g) + h = f + (g + h)$. Now for $x \in X$

$$
\begin{aligned}
[(f + g) + h](x) &= (f + g)(x) + h(x) & \text{(definition)} \\
&= [f(x) + g(x)] + h(x) & \text{(definition)} \\
&= f(x) + [g(x) + h(x)] & \text{(associativity in } \mathbb{R}) \\
&= f(x) + (g + h)(x) & \text{(definition)} \\
&= [f + (g + h)](x) & \text{(definition)}.
\end{aligned}
$$

Hence, by the condition for the equality of mappings

$$(f + g) + h = f + (g + h).$$

1.3 The zero of R is given by the mapping 0 given by $0(x) = 0$ $(x \in X)$ since for all $f \in R$ and $x \in X$

$$
\begin{aligned}
(f + 0)(x) &= f(x) + 0(x) && \text{(definition)} \\
&= f(x) + 0 && \text{(definition of } 0 \in R) \\
&= f(x) && (0 \in \mathbb{R}) \\
&= 0 + f(x) && (0 \in \mathbb{R}) \\
&= 0(x) + f(x) && \text{(definition of } 0 \in R) \\
&= (0 + f)(x) && \text{(definition).}
\end{aligned}
$$

Hence, again by the condition for the equality of mappings,

$$f + 0 = f = 0 + f.$$

1.4 The additive inverse of $f \in R$ is given by $-f$ where $(-f)(x) = -f(x)$ $(x \in X)$. This is well-defined since on the right-hand side we have the negative of a real number. We must verify that $-f + f = 0 = f + (-f)$. Now for $x \in X$

$$
\begin{aligned}
(-f + f)(x) &= (-f)(x) + f(x) && \text{(definition)} \\
&= -f(x) + f(x) && \text{(definition of } -f) \\
&= 0 \\
&= 0(x) && \text{(definition of } 0 \in R).
\end{aligned}
$$

Thus

$$-f + f = 0.$$

Similarly $f + (-f) = 0$.

1.5 Let $f, g \in R$. Then for $x \in X$

$$
\begin{aligned}
(f + g)(x) &= f(x) + g(x) && \text{(definition)} \\
&= g(x) + f(x) && \text{(commutativity of addition in } \mathbb{R}) \\
&= (g + f)(x) && \text{(definition).}
\end{aligned}
$$

Thus

$$f + g = g + f$$

and so addition in R is commutative.

2.1 By definition R is closed under multiplication.

2.2 Let $f, g, h \in R$. We want to show that $(f.g).h = f.(g.h)$. Now for $x \in X$

$$[(f.g).h](x) = (f.g)(x)h(x) \quad \text{(definition)}$$
$$\cdot \quad = [f(x)g(x)]h(x) \quad \text{(definition)}$$
$$= f(x)[g(x)h(x)] \quad \text{(associativity in } \mathbb{R})$$
$$= f(x)(g.h)(x) \quad \text{(definition)}$$
$$= [f.(g.h)](x) \quad \text{(definition)}.$$

Thus

$$(f.g).h = f.(g.h).$$

3.1 Let $f, g, h \in R$. We prove that $f.(g+h) = f.g + f.h$. Let $x \in X$. Then

$$[f.(g+h)](x) = f(x)(g+h)(x) \quad \text{(definition)}$$
$$= f(x)[g(x) + h(x)] \quad \text{(definition)}$$
$$= f(x)g(x) + f(x)h(x) \quad \text{(distributivity in } \mathbb{R})$$
$$= (f.g)(x) + (f.h)(x) \quad \text{(definition)}$$
$$= [(f.g) + (f.h)](x) \quad \text{(definition)}.$$

Thus

$$f.(g+h) = f.g + f.h.$$

The other distributivity axiom is proved similarly. We have now shown that R is a ring.

We may remark, partly as an aside, that as the ring of integers $\mathbb{Z}$ is an example of a system that satisfies the axioms of a ring then these axioms are necessarily self-consistent. We shall not concern ourselves unduly in determining to what extent the axioms of this, or of any other, system are independent of one another.

We have not, however, adequately considered all of the distinctive features of the ring of integers and of the ring of polynomials. In $\mathbb{Z}$ there are three properties which are so familiar that ordinarily we do not make specific reference to them: there is a number 1 such that $1a = a1 = a$ for all $a \in \mathbb{Z}$, the order in which we multiply integers is irrelevant ($ab = ba$ for all $a, b \in \mathbb{Z}$) and the product of two non-zero integers is non-zero. We formalize these concepts as follows.

Definition 5

Let R be a ring.

1. An element $1 \in R$ such that $1a = a1 = a$ for all $a \in R$ is called an **identity element** or **identity** or **unity** of R.

2. Let a and b be elements of R. Then a and b are said to **commute** if the products ab and ba are equal. If any two elements of R commute then R is said to be **commutative**.

3. Let 0 be the zero element of R. Let a and b be elements of R such that $ab = 0$. Then a and b are called **divisors of zero** and if both $a \neq 0$ and $b \neq 0$ then a and b are called **proper divisors of zero**.

We have previously shown above that the zero element of a ring is unique. By an entirely similar argument we may show that if a ring has an identity element (which need not be the case) then it has a unique identity element.

Definition 6

A commutative ring with an identity $1(1 \neq 0)$ and no proper divisors of zero is called an **integral domain**.

(Note that the use here of the word 'domain' in the phrase 'integral domain' has no connection with the 'domain' we encountered previously in the mapping of one set into another. Note also that some texts, especially of an advanced nature, do not insist upon 'commutativity' as a necessary condition for an integral domain.)

Example 9

Under the usual operations of addition and multiplication the following are all integral domains,

$$\mathbb{Z}, \mathbb{Q}, \mathbb{R}, \mathbb{C}.$$

As part of this introduction to rings we require to prove some elementary results, results which in themselves are obvious for the integers but which require proof in an axiomatic context.

Theorem 3

Let R be a ring. Then the following hold.

1. If $a + b = a + c$ $(a, b, c \in R)$ then $b = c$.

   If $a + b = c + b$ $(a, b, c \in R)$ then $a = c$.

2. For all $a \in R$, $a0 = 0a = 0$.

3. For all $a, b \in R$, $a(-b) = (-a)b = -(ab)$, $(-a)(-b) = ab$.
   If R has an identity 1 then $(-1)(-1) = 1$.
4. For all $a, b \in R$, $a(b - c) = ab - ac$, $(a - b)c = ac - bc$.

Proof

1. From $a + b = a + c$ we have $(-a) + (a + b) = (-a) + (a + c)$. By associativity

$$((-a) + a) + b = ((-a) + a) + c$$

and so

$$0 + b = 0 + c \text{ and } b = c.$$

The second assertion is similarly proved.

2. $a0 + a0 = a(0 + 0) = a0 = a0 + 0$ and so $a0 = 0$. Similarly $0a = 0$.
3. We have $0 = a0 = a(b + (-b)) = ab + a(-b)$ and so $ab + -(ab) = 0 = ab + a(-b)$ from which $-(ab) = a(-b)$. Similarly we prove $-(ab) = (-a)b$.
   Let $a' = -a$. Then, as we observed in the Commentary above, $-a' = -(-a) = a$. Hence

$$(-a)(-b) = a'(-b) = (-a')b = ab.$$

If $1 \in R$ then $(-1)(-1) = 11 = 1$.

4. $a(b - c) = a(b + (-c)) = ab + a(-c) = ab + -(ac) = ab - ac$. Similarly

$$(a - b)c = ac - bc. \qquad \square$$

Example 10

Let D be an integral domain. Let a, b, $c \in C$, $c \neq 0$, be such that $ac = bc$. We want to prove that $a = b$. We have $(a - b)c = ac - bc = 0$ and since D is an integral domain either $a - b = 0$ or $c = 0$. But $c \neq 0$ and so $a - b = 0$ and $a = b$.

We have considered polynomial rings in x in which the polynomials have coefficients from $\mathbb{Z}$, $\mathbb{Q}$ or $\mathbb{R}$, yielding $\mathbb{Z}[x]$, $\mathbb{Q}[x]$ or $\mathbb{R}[x]$ respectively. But similarly we may have a polynomial ring in x in which the coefficients belong to an integral domain D yielding $D[x]$. We have the following result.

Theorem 4

Let D be an integral domain. Then the polynomial ring $D[x]$ is an integral domain.

Proof

Certainly $D[x]$ is a commutative ring with the identity of D as the identity of $D[x]$. Thus we have to show that $D[x]$ has no proper divisors of zero. Let

$$f(x) = a_0 + a_1 x + \ldots + a_m x^m \quad (a_m \neq 0),$$
$$g(x) = b_0 + b_1 x + \ldots + b_n x^n \quad (b_n \neq 0),$$

be two non-zero polynomials in $D[x]$. Then $f(x)g(x)$ is a polynomial of highest term $a_m b_n x^{m+n}$. But since D is an integral domain $a_m b_n \neq 0$. Thus $f(x)g(x) \neq 0$ and $D[x]$ is an integral domain. $\qquad \square$

Corollary

Let $f(x)$ and $g(x)$ be non-zero polynomials in $D[x]$. Then

$$\deg f(x)g(x) = \deg f(x) + \deg g(x).$$

In many situations we have to consider a ring which is itself contained within a larger ring, for example $\mathbb{Z}$ is a ring within the ring $\mathbb{Q}$ and $\mathbb{Q}$ is a ring within the ring $\mathbb{R}$. We sometimes have to consider subsets of a ring which may be rings, or 'subrings' as they will be called, with respect to the operations of the ring.

Definition 7

Let S be a non-empty subset of the ring R which is also a ring under the addition and multiplication in R. Then S is called a **subring** of R.

This definition is too descriptive to be really useful. We therefore derive a criterion which is more readily applicable.

Theorem 5 Subring Criterion

A non-empty subset S of a ring R is a subring of R if and only if the following axioms of a ring are satisfied.

1.1 S is closed under addition.

1.4 For all $a \in S$, $-a \in S$.

2.1 S is closed under multiplication.

Proof

Certainly if S is a subring the axioms quoted above must be satisfied.

Suppose now that the axioms above are satisfied by S. Then certainly we have $0 = a + (-a) \in S$. We claim that the remaining axioms for a ring are automatically satisfied in S because they are satisfied in R. Consider, for the sake of argument, the associativity of addition. We have $(a + b) + c = a + (b + c)$ for all a, b, $c \in R$ and therefore for all a, b, $c \in S$ as $S \subseteq R$. The other remaining axioms for a ring are similarly satisfied since $S \subseteq R$. Hence we infer that S is a subring. $\square$

Example 11

It may be shown that $R = M_2(\mathbb{Q})$, the set of 2×2 matrices over $\mathbb{Q}$, is a ring under the usual matrix operations. Suppose we want to show that

$$S = \left\{ \begin{pmatrix} x & y \\ 0 & z \end{pmatrix} \mid x, y, z \in \mathbb{Q} \right\}$$

is a subring of R. We apply the subring criterion. We have, for $x, y, z, u, v, w \in \mathbb{Q}$,

$$\begin{pmatrix} x & y \\ 0 & z \end{pmatrix} + \begin{pmatrix} u & v \\ 0 & w \end{pmatrix} = \begin{pmatrix} x+y, & y+r \\ 0 & z+w \end{pmatrix} \in S,$$

$$-\begin{pmatrix} x & y \\ 0 & z \end{pmatrix} = \begin{pmatrix} -x & -y \\ 0 & -z \end{pmatrix} \in S,$$

$$\begin{pmatrix} x & y \\ 0 & z \end{pmatrix}\begin{pmatrix} u & v \\ 0 & w \end{pmatrix} = \begin{pmatrix} xu & xv+yw \\ 0 & zw \end{pmatrix} \in S.$$

where we are using the closure etc. of addition and multiplication in $\mathbb{Q}$ to ensure that, for example, $\begin{pmatrix} x+y & y+v \\ 0 & z+w \end{pmatrix}$ is indeed in S.

We conclude that S is a subring of R.

Exercises 3.3

1. Prove that if a ring has an identity then it has only one identity.

2. Verify that $\mathbb{Z}$, $\mathbb{Q}$, $\mathbb{R}$ and $\mathbb{C}$ with the usual operations of addition and multiplication are integral domains.

3. Show that the set R of one element 0, $R = \{0\}$, is a ring if addition and multiplication are given by $0 + 0 = 0$ and $00 = 0$ respectively.

4. Let R be a set of two elements, 0 and 1, $R = \{0, 1\}$, in which addition and multiplication are given by the following tables:

Sum	0	1
0	0	1
1	1	0

Product	0	1
0	0	0
1	0	1

Prove that R is an integral domain.

5. Let $2\mathbb{Z}$ denote the set of even integers with the usual operations of addition and multiplication. Prove that $2\mathbb{Z}$ is a ring. Is $2\mathbb{Z}$ an integral domain?

6. Let R be a ring.

(i) Let $a, b, c \in R$ be such that $a + b = c + b$. Prove that $a = c$.

(ii) Let $a, b \in R$. Prove that $-(ab) = (-a)b$.

7. Let $M_2(\mathbb{Q})$ be the set of 2×2 matrices over $\mathbb{Q}$. Under the usual operations prove that $M_2(\mathbb{Q})$ is a ring.

8. Let R_1 and R_2 be subrings of a ring R. Prove that $R_1 \cap R_2$ is a subring of R. Can you generalize this result?

4
Introduction to Groups

In Chapter 3, we began to consider an axiomatic treatment of certain algebraic concepts and operations; in particular in the case of a ring we needed to consider two binary operations. Since simplicity usually has some advantages we shall in this chapter consider only one binary operation. We begin therefore with the notion of a 'semigroup' from which we shall be led to consider a 'group', one of the most fundamental structures in modern mathematics.

The evolution of the concept of an abstract group owes much to the labours of many mathematicians of whom only a few will be mentioned here. The origins of the concept may be traced from the work of P. Ruffini (1765–1822) and E. Galois (1811–32) through to that of L. Kronecker who developed ideas for what we now call an Abelian group ('Abelian' after N.H. Abel, 1802–29). The abstract concept of a finite group was first formulated in 1854 by A. Cayley (1821–95) but its significance was not properly appreciated until 1878. W. von Dyck (1856–1934) and H. Weber (1842–1913) were influential in the development of group theory, the latter giving the first definition of an infinite group in 1893. These remarks can give at most a brief indication of the history of the evolution of the concept of an abstract group which "emerged not from a single act or the creation of a single scholar but was rather the outcome of a process of abstraction with certain discernible steps. These were not essentially logical in nature; rather they represented the variable extent to which the process of abstraction was carried through." (quotation from The Genesis of the Abstract Group Concept by H. Wussing, translated MIT Press, 1984, page 234). It is perhaps psychologically salutary to reflect that the concepts of a group and, indeed, of other parts of mathematics, have not always evolved directly but frequently by fits and starts.

4.1 Semigroups

Definition 1

Let S be a non-empty set on which there is defined a binary operation denoted by
$*$. For $a, b \in S$ the outcome of the operation between a and b is denoted by $a * b$.
Then S is called a **semigroup** if the following axioms hold.

(i) For all $a, b \in S$, $a * b \in S$ (closure).
(ii) For all $a, b, c \in S$, $(a * b) * c = a * (b * c)$ (associativity).

Example 1

$\mathbb{Z}$ under the operation of $+$ is a semigroup since for all $a, b \in \mathbb{Z}$, $a + b \in \mathbb{Z}$ and for
all $a, b, c \in \mathbb{Z}$, $(a + b) + c = a + (b + c)$.

In future, in general considerations of semigroups, we shall omit the symbol for
the binary operation unless an especial need arises for it to be shown. We shall
use, as far as possible, a multiplicative notation in which the outcome of the
binary operation between two elements is shown by the simple juxtaposition
of the elements; we shall speak of the **multiplication** (sometimes **addition**) of
these elements.

Semigroups are common in mathematics and easy to create as the following
examples show.

Examples 2

1. Let S be a set of one element, $S = \{a\}$, say. Define

$$aa = a.$$

   Then S becomes a semigroup.

2. Let S be a non-empty set. Define a binary operation on S by

$$ab = b \text{ for all } a, b \in S.$$

   We claim that S is a semigroup under this operation. Certainly we have for
   all $a, b \in S$, $ab = b \in S$. We also have for all $a, b, c \in S$, $(ab)c = bc = c$ and
   $a(bc) = ac = c$ and so $(ab)c = a(bc)$. Thus S is a semigroup.

3. Let S be a non-empty set and let s be a fixed element of S. Define a binary
   operation on S by

$$ab = s \text{ for all } a, b \in S.$$

Then certainly for all $a, b \in S$, $ab \in S$ and then we have for all $a, b, c \in S$, $(ab)c = sc = s$ and $a(bc) = as = s$ from which $(ab)c = a(bc)$. Thus S is a semigroup.

None of the examples above may be of great intrinsic interest but nevertheless in each example the axioms of a semigroup are satisfied. After the next lemma we consider a particular type of semigroup.

Lemma 1

Let S be a semigroup. Let $e, f \in S$ be such that $ex = x$ and $xf = x$ for all $x \in S$. Then $e = f$ and so $ex = xe = x$ for all $x \in S$.

Proof

By the condition on e we have, in particular, $ef = f$. Similarly $ef = e$. Thus $e = f$ and the rest is clear.   □

Definition 2

A semigroup S is called a **monoid** if there exists $e \in S$ such that $ex = xe = x$ for all $x \in S$. e is said to be the **identity element** or **identity** of S.

By Lemma 1 above a semigroup has at most one identity element.

Associativity and Index Law

Let $a, b, c, d, \ldots$ be elements of a semigroup S. Expressions such as $(ab)(cd)$, $a((bc)d)$, $a(b(cd))$ are formally distinct but are in fact equal since, by associativity,

$$(ab)(cd) = a(b(cd)) = a((bc)d).$$

Thus we could write, without uncertainty of meaning, simply $abcd$. Further we see that in more complicated expressions of elements and brackets we may similarly dispense with the bracketing provided the elements appear in the same order in each of the expressions.

Consequently we may now define the powers of $a \in S$ unambiguously as

$$a^1 = a, a^2 = aa, a^3 = aaa, \ldots, a^n = aa \ldots a,$$

where a appears precisely n times. It follows that for $m, n = 1, 2, \ldots$

$$a^m a^n = (aa \ldots a)(aa \ldots a) = (aa \ldots a) = a^{m+n}$$

where within the brackets we have a appearing m times, n times and $m + n$ times respectively.

If further S is a monoid with identity e we may define $a^0 = e$ for all $a \in S$ and then

$$a^m a^n = a^{m+n} \quad (m, n = 0, 1, \ldots).$$

Examples 3

1. Let X be a non-empty set and let $S(X)$ be the set of mappings of X into X. We have already met the circle composition of mappings of X into X, namely for $f, g \in S(X)$ we defined $f \circ g$ by

$$(f \circ g)(x) = f(g(x)) \quad (x \in X).$$

Furthermore we showed (Theorem 7, Chapter 1) that for $f, g, h \in S(X)$ then

$$(f \circ g) \circ h = f \circ (g \circ h).$$

The set $S(X)$ of mappings of X into X is therefore a semigroup under the circle composition of mappings. $S(X)$ has an identity element given by the mapping ι where $\iota(x) = x (x \in X)$ since for all $f \in S(X)$ we have

$$(\iota \circ f)(x) = \iota(f(x)) = f(x) = f(\iota(x)) = (f \circ \iota)(x) \quad (x \in X)$$

and so

$$\iota \circ f = f = f \circ \iota.$$

2. Let a binary operation $*$ be introduced into $\mathbb{Z}$ by defining for all $a, b \in \mathbb{Z}$

$$a * b = a + b + ab.$$

Then we claim that $\mathbb{Z}$ with this operation is a monoid. We certainly have $a * b \in \mathbb{Z}$ for all $a, b \in \mathbb{Z}$. Let $a, b, c \in \mathbb{Z}$, then

$$\begin{aligned}
(a * b) * c &= (a + b + ab) * c \\
&= (a + b + ab) + c + (a + b + ab)c \\
&= a + b + ab + c + ac + bc + abc \\
&= a + b + c + ab + ac + bc + abc,
\end{aligned}$$

$$a * (b * c) = a * (b + c + bc)$$
$$= a + (b + c + bc) + a(b + c + bc)$$
$$= a + b + c + bc + ab + ac + abc$$
$$= a + b + c + ab + ac + bc + abc,$$

from which we have

$$(a * b) * c = a * (b * c).$$

Now for all $a \in \mathbb{Z}$ we have

$$0 * a = 0 + a + 0a = 0 + a + 0 = a,$$
$$a * 0 = a + 0 + a0 = a + 0 + 0 = a.$$

Thus 0 plays the role of an identity element when $*$ is the binary operation. Hence $\mathbb{Z}$ is a monoid under $*$.

3. The set $M_2(\mathbb{Q})$ of 2×2 matrices with entries from $\mathbb{Q}$ is a monoid under the usual matrix addition, namely

$$\begin{pmatrix} x_1 & y_1 \\ z_1 & t_1 \end{pmatrix} + \begin{pmatrix} x_2 & y_2 \\ z_2 & t_2 \end{pmatrix} = \begin{pmatrix} x_1 + x_2 & y_1 + y_2 \\ z_1 + z_2 & t_1 + t_2 \end{pmatrix}.$$

The sum of two 2×2 matrices with entries from $\mathbb{Q}$ is again a 2×2 matrix over $\mathbb{Q}$ and so the operation of addition is closed. The associativity of addition is a well known property of matrix addition. Further

$$\begin{pmatrix} 0 & 0 \\ 0 & 0 \end{pmatrix}$$

is the identity element for addition and so $M_2(\mathbb{Q})$ is a monoid.

4. The set $M_2(\mathbb{Q})$ of 2×2 matrices with entries from $\mathbb{Q}$ is a monoid under the usual matrix multiplication, namely

$$\begin{pmatrix} x_1 & y_1 \\ z_1 & t_1 \end{pmatrix} + \begin{pmatrix} x_2 & y_2 \\ z_2 & t_2 \end{pmatrix} = \begin{pmatrix} x_1 x_2 + y_1 z_2 & x_1 y_2 + y_1 t_2 \\ z_1 x_2 + t_1 z_2 & z_1 y_2 + t_1 t_2 \end{pmatrix}.$$

Closure is obvious and matrix multiplication is known to be associative. The identity element for multiplication is

$$\begin{pmatrix} 1 & 0 \\ 0 & 1 \end{pmatrix}$$

We conclude this section with two results which will be important in our study of groups.

Lemma 2

Let S be a monoid with identity element e. Let $a \in S$. Let a', $a'' \in S$ be such that $a'a = e = aa''$. Then $a' = a''$.

Proof

The result depends crucially on the associativity axiom. We have

$$(a'a)a'' = ea'' = a'', \quad a'(aa'') = a'e = a' \quad \text{and so} \quad a' = a''. \qquad \square$$

Theorem 1

Let S be a semigroup with the following two conditions.

1. There exists $e \in S$ such that $ec = a$ for all $a \in S$.
2. For each $a \in S$ there exists $a' \in S$ such that $a'a = e$.

   Then S is a monoid with identity e and $a'a = aa' = e$.

Proof

We show first that e is the identity of S. Let $a \in S$. Then, by twice applying condition 2, there exists $a' \in S$ such that $a'a = e$ and there exists $a'' \in S$ such that $a''a' = e$.

But by condition 1 and associativity we have

$$ae = e(ae) = (a''a')(ae) = a''(a'(ae)) = a''((a'a)e)$$

$$= a''(ee) = a''e = a''(a'a) = (a''a')a = ea = a.$$

Consequently e is the identity of S. But then $a'a = e = a''a'$ implies, by Lemma 1, that $a = a''$ and so we complete the proof. $\qquad \square$

We remark that in the two conditions of this theorem e'acts' on the left of any $a \in S$ to give $ea = a$ and for given $a \in S$ a' 'acts' on the left to give $a'a = e$. If we were to replace these left-handed conditions by corresponding right-handed conditions then the conclusion of the theorem would still hold. If, however, one condition is on the left and the other condition is on the right then the conclusion of the theorem need not hold. To see this last assertion we consider the following example.

Example 4

Let S be the two-element semigroup, $S = \{e, f\}$ with multiplication, $ee = e$, $ef = f$, $ff = f$, $fe = e$. S is a semigroup of the type described in Examples 2 no. 2, where e acts on the left and satisfies condition 1 of the last theorem. We also have $ee = e$ and $fe = e$ so that, from the right we would have, $e' = e$ and $f' = e$. Nevertheless e is not an identify element of S.

Exercises 4.1

1. A binary operation $*$ is defined on $\mathbb{Z}$ by $a * b = ba$ $(a, b \in \mathbb{Z})$. Prove that, under this operation, $\mathbb{Z}$ is a monoid.

2. A mapping $f_{a,b}$ of $\mathbb{R}$ into $\mathbb{R}$ is defined for $a, b \in \mathbb{R}$ by

$$f_{a,b} : x \to a + bx \quad (x \in \mathbb{R}).$$

Prove that the set of all such mappings is a monoid under the circle composition of mappings. Hint: Prove that

$$f_{c,d} \circ f_{a,b} = f_{c+ad,bd}.$$

3. Let t be a given integer. On $\mathbb{Z}$ an operation $*$ is defined by

$$a * b = a + b + tab \quad (a, b \in \mathbb{Z}).$$

Prove that $\mathbb{Z}$ under this operation is a monoid.

4. Let S be a semigroup with the following two conditions.
   (i) There exists $e \in S$ such that $ae = a$ for all $a \in S$.
   (ii) For all $a \in S$ there exists $a' \in S$ such that $aa' = e$.
   Prove that e is the identify of S and that $a'a = e$.

5. (Hard) Let S be a finite semigroup and let $a \in S$. By considering the subset $\{a^m \,|\, n \in \mathbb{N}\}$ prove that there exist $p, q \in \mathbb{N}$, $2q < p$ such that $a^p = a^q$. If $b = a^{p-q}$ prove that $b^2 = b$.

4.2 Finite and Infinite Groups

Definition 3

Let G be a non-empty set on which there is defined a binary operation so that the outcome of the operation between a and b $(a, b \in G)$ is denoted by ab. Then G is called a **group** if the following axioms hold.

1. For all $a, b \in G$, $ab \in G$ (closure).
2. For all $a, b, c \in G$, $(ab)c = a(bc)$ (associativity).
3. There exists $e \in G$ such that for all $a \in G$, $ea = a$ (existence of identity).
4. For each $a \in G$ there exists an element denoted by a^{-1} such that $a^{-1}a = e$ (existence of inverse).

By Theorem 1 e **is** the **identity** of G and, for each $a \in G$, we have a^{-1} uniquely determined by a and such that, $a^{-1}a = aa^{-1} = e$, a^{-1} is called the **inverse** of a.

Notice that from $a^{-1}a = aa^{-1} = e$ we may deduce that

$$(a^{-1})^{-1} = a \quad (a \in G).$$

Partly for later convenience we give two alternative criteria by which a semi-group is a group.

Theorem 2

Let S be a semigroup. Suppose that for any $a, b \in S$ there exist $x, y \in S$ such that $ax = b$ and $ya = b$. Then S is a group.

Proof

Let $a \in S$. By assumption there exists $e \in S$ such that $ea = a$. We cannot imme-diately assert that e is the identity for S since, on the face of it, this e depends on the particular element a. We consider an arbitrary $c \in S$. Then again by assump-tion there exists $u \in S$ such that $au = c$. Hence $ec = e(au) = (ea)u = au = c$. Furthermore by assumption there exists $c' \in S$ such that $c'c = e$. Hence, finally, S is a group. □

Theorem 3

Let S be a finite semigroup in which cancellation exists, that is if $a, b, x \in S$ and $ax = bx$ then $a = b$ and if $a, b, y \in S$ and $ya = yb$ then $a = b$. Then S is a finite group.

Proof

Let $a, b \in S$. We show that there exists $x \in S$ such that $ax = b$. Let S have n dis-tinct elements $s_1, s_2, \ldots, s_n$. Consider $as_1, as_2, \ldots, as_n$. These elements of S are also distinct since $as_i = as_j$ implies $s_i = s_j$ and so $i = j$. Thus

$$\{as_1, as_2, \ldots as_n\} = \{s_1, s_2, \ldots, s_n\}.$$

Hence, as b is one of the $s_1, s_2, \ldots, s_n$, there exists s_k such that $as_k = b$. Similarly we may find $y \in S$ such that $ya = b$. Thus, by the previous result, S is a group. $\square$

Since the group concept is widely encountered in mathematics we shall give a varied selection of examples, not all of which need to be fully comprehended on a first reading. For convenience in this chapter, and elsewhere, we shall use G to denote a group and e to denote the identity of G without further explanation.

Examples 5

1. The set of non-zero elements of $\mathbb{Q}$, that is $\mathbb{Q} \setminus \{0\}$, is a group under the usual multiplication. To verify that $\mathbb{Q} \setminus \{0\}$ is a group we note the following immediate facts. If $a, b \in \mathbb{Q} \setminus \{0\}$ then $ab \in \mathbb{Q} \setminus \{0\}$. Multiplication in $\mathbb{Q}$, and so in $\mathbb{Q} \setminus \{0\}$, is associative. 1 is the identity as $1a = a1 = a$ for all $a \in \mathbb{Q} \setminus \{0\}$. If finally $a \in \mathbb{Q} \setminus \{0\}$ then a has a reciprocal $\dfrac{1}{a}$ such that $\left(\dfrac{1}{a}\right)a = a\left(\dfrac{1}{a}\right) = 1$ and so $a^{-1} = \dfrac{1}{a}$.

   In a similar manner we may prove that $\mathbb{R} \setminus \{0\}$ and $\mathbb{C} \setminus \{0\}$ are groups.

2. The set of integers $\mathbb{Z}$ under multiplication is not a group. We know that this set is a monoid under multiplication but, for example, the number 2 does not have an inverse in $\mathbb{Z}$ as $\frac{1}{2} \in \mathbb{Z}$.

3. The set of integers $\mathbb{Z}$ is a group under the usual addition. Here we have to insert the specific symbol $+$ for the binary operation but the group axioms are easily verified. Thus if $a, b \in \mathbb{Z}$ then $a + b \in \mathbb{Z}$ and for all $a, b, c \in \mathbb{Z}$

$$(a + b) + c = a + (b + c).$$

The element 0 plays the role of the identity element for addition since

$$0 + a = a + 0 = a$$

for all $a \in \mathbb{Z}$ and the inverse of $a \in \mathbb{Z}$ with respect to addition is $-a$, since

$$(-a) + a = a + (-a) = 0.$$

We now gather together some elementary results on groups.

Theorem 4

Let G be a group.

1. **Cancellation**
   Let $a, b, x \in G$ be such that $ax = bx$. Then $a = b$.
   (Similarly $ya = yb$ $(y \in G)$ implies $a = b$.)

2. **Unique Solution of Equation**

   Let $a, b \in G$. Then the equation $ax = b$ has the unique solution $x = a^{-1}b$.
   (Similarly $ya = b$ has the unique solution $y = ba^{-1}$.)

Proof

1. There exists $x^{-1} \in G$ such that $xx^{-1} = e$. Then $(ax)x^{-1} = (bx)x^{-1}$ and so $a(xx^{-1}) = b(xx^{-1})$ from which $ae = be$ and $a = b$.
2. Certainly $a^{-1}b$ is a solution since $a(a^{-1}b) = (aa^{-1})b = eb = b$.

   On the other hand, $ax = b$ implies that $a^{-1}(ax) = a^{-1}b$ from which we conclude that $(a^{-1}a)x = a^{-1}b$ and so $x = ex = a^{-1}b$. $\square$

Definition 4

A group G is said to be **finite** if the set G is a finite set, otherwise the group is said to be **infinite**. If G is finite the cardinality of G (which is the number of elements in G) is called the **order** of G, written $|G|$, G then has **finite order**. An infinite group is often said to have **infinite order**.

Example 6

$\mathbb{Z}$ under addition and $\mathbb{Q} \setminus \{0\}$ under multiplication are examples of groups of infinite order.

In the case of a finite group of very small order we may conveniently display the binary operation in the tabular form of a so-called 'Cayley table'. Thus if G has order n, $G = \{a_1, a_2, \ldots, a_n\}$ say, then we write the table as

	a_1	a_2	$\ldots$	a_j	$\ldots$	a_n
a_1	a_1a_1	a_1a_2	$\ldots$	a_1a_j	$\ldots$	a_1a_n
a_2	a_2a_1	a_2a_2	$\ldots$	a_2a_j	$\ldots$	a_2a_n
$\ldots$	$\ldots$	$\ldots$	$\ldots$	$\ldots$	$\ldots$	$\ldots$
a_i	a_ia_1	a_ia_2	$\ldots$	a_ia_j	$\ldots$	a_ia_n
$\ldots$	$\ldots$	$\ldots$	$\ldots$	$\ldots$	$\ldots$	$\ldots$
a_n	a_na_1	a_na_2	$\ldots$	a_na_j	$\ldots$	a_na_n

where the entry in the (i, j)th position is a_ia_j. By the cancellation result in Theorem 4 the elements $a_ia_1, a_ia_2, \ldots, a_ia_n$ are distinct and so any row of the Cayley table must have distinct elements in the row. Similarly any column of the table must consist of distinct elements.

Examples 7

1. Let G have order 1. Then $G = \{e\}$ and the Cayley table is

$$
\begin{array}{c|c}
 & e \\
\hline
e & e
\end{array}
$$

As a concrete example of this group we have $\{1\}$ as a multiplicative group.

2. Let G have order 2. Then $G = \{e, a\}$, say. Now $a^2 = a$ or e and $a^2 = a$ implies $aa = a^2 = ae$ and so $a = e$ which is false. Thus $a^2 = e$. The Cayley table is

$$
\begin{array}{c|cc}
 & e & a \\
\hline
e & e & a \\
a & a & e
\end{array}
$$

As a concrete example we have $\{1, -1\}$ under multiplication.

3. Let G have order 3, say, $G = \{e, a, b\}$. Consider ab. If $ab = a$ or $ab = b$ we have, by cancellation, that $b = e$ or $a = e$ respectively and either inference is false. Thus $ab = e$. Hence $b = a^{-1}$ and so also $ba = e$. We obtain the partially completed Cayley table

$$
\begin{array}{c|ccc}
 & e & a & b \\
\hline
e & e & a & b \\
a & a & & e \\
b & b & e &
\end{array}
$$

But each row and column must contain distinct elements and so the Cayley table is

$$
\begin{array}{c|ccc}
 & e & a & b \\
\hline
e & e & a & b \\
a & a & b & e \\
b & b & e & a
\end{array}
$$

The set of matrices

$$
\left\{ \begin{pmatrix} 1 & 0 & 0 \\ 0 & 1 & 0 \\ 0 & 0 & 1 \end{pmatrix}, \begin{pmatrix} 0 & 1 & 0 \\ 0 & 0 & 1 \\ 1 & 0 & 0 \end{pmatrix}, \begin{pmatrix} 0 & 0 & 1 \\ 1 & 0 & 0 \\ 0 & 1 & 0 \end{pmatrix} \right\}
$$

forms a group with the same table if we put

$$e = \begin{pmatrix} 1 & 0 & 0 \\ 0 & 1 & 0 \\ 0 & 0 & 1 \end{pmatrix}, a = \begin{pmatrix} 0 & 1 & 0 \\ 0 & 0 & 1 \\ 1 & 0 & 0 \end{pmatrix}, b = \begin{pmatrix} 0 & 0 & 1 \\ 1 & 0 & 0 \\ 0 & 1 & 0 \end{pmatrix}$$

4. Let G have order 4. Then $G = \{e, a, b, c\}$, say. We must distinguish two cases.

(i) Suppose there exists an element in G which is not its own inverse and let a be such an element. Thus $a^{-1} \neq a$. For the sake of argument suppose $a^{-1} = b$. Thus $ab = ba = e$. But then $ac \neq a$ and $ac \neq c$, as otherwise $c = e$ or $a = e$ respectively, hence we must have $ac = b$. Similarly $ca \neq e, a, c$ and so $ca = b$. We now obtain the incomplete Cayley table

	e	a	b	c
e	e	a	b	c
a	a		e	b
b	b	e		
c	c	b		

But the elements in the second row must necessarily be distinct and so we have $a^2 = c$. Also $ac = b$ implies $bac = b^2$ and so, as $ba = e$, we obtain $c = b^2$. Thus maintaining the distinctiveness of the elements in any given row or column we obtain the Cayley table

	e	a	b	c
e	e	a	b	c
a	a	c	e	b
b	b	e	c	a
c	c	b	a	e

A possible realization is the group $\{1, i, -1, -i\}$.

(ii) Suppose now that every element is its own inverse. Then $a^2 = b^2 = c^2 = e$. Then $ab \neq e, a, b$ and so $ab = c$, also $ba \neq e, a, b$ and so $ba = c$. Similarly we deduce that $ac = ca = b$ and $bc = cb = a$. The Cayley table is

	e	a	b	c
e	e	a	b	c
a	a	e	c	b
b	b	c	e	a
c	c	b	a	e

This abstract group is known as the **Klein four-group** (after F. Klein, 1849–1925). A realization is afforded by a group of four matrices, namely

$$\left\{ \begin{pmatrix} 1 & 0 \\ 0 & 1 \end{pmatrix}, \begin{pmatrix} -1 & 0 \\ 0 & -1 \end{pmatrix}, \begin{pmatrix} 0 & 1 \\ 1 & 0 \end{pmatrix}, \begin{pmatrix} 0 & -1 \\ -1 & 0 \end{pmatrix} \right\}.$$

Theorem 5

Let $a, b \in G$. Then

$$(ab)^{-1} = b^{-1}a^{-1}.$$

Proof

Inserting brackets for clarity we have

$$(b^{-1}a^{-1})(ab) = b^{-1}(a^{-1}(ab)) = b^{-1}((a^{-1}a)b) = b^{-1}(eb) = b^{-1}(b) = e.$$

Hence

$$b^{-1}a^{-1} = (ab)^{-1}.$$

□

Corollary

Let $a_1, a_2, \ldots, a_n \in G$. Then $(a_1 a_2 \ldots a_n)^{-1} = a_n^{-1} a_{n-1}^{-1} \ldots a_1^{-1}$.

Proof

A simple induction suffices.

□

We now wish to extend the index law for powers of $a \in G$, namely

$$a^m a^n = a^{m+n} \quad (m, n = 0, 1, \ldots)$$

to the case in which m and n may be negative. First we have to define a^{-n} $(n > 0)$.

Definition 5

Let $a \in G$. Let $n \in \mathbb{N}$. Then a^{-n} is defined by

$$a^{-n} = (a^{-1})^n$$

Using this definition, which is entirely natural if we think of reciprocals of rational numbers, we may extend the index law above. We give examples which show how the extension is used.

Example 8

Let $a \in G$.

$$(a^3)(a^{-1})^2 = (aaa)(a^{-1}a^{-1}) = aa(aa^{-1})a^{-1}$$

$$= aaea^{-1} = aaa^{-1} = a(aa^{-1}) = ae = a.$$

$$(a^{-3})(a)^2 = (a^{-1}a^{-1}a^{-1})(aa) = a^{-1}a^{-1}(a^{-1}a)a = a^{-1}a^{-1}(e)a$$

$$= a^{-1}a^{-1}a = a^{-1}(a^{-1}a) = a^{-1}e = a^{-1}.$$

Theorem 6 Index Law for Groups

Let $a \in G$. Then

$$a^m a^n = a^{m+n} \quad (m, n \in \mathbb{Z}).$$

Proof (may be omitted on a first reading)

We may clearly confine our argument to the cases in which one, or both, of m and n is negative. We shall only consider here the case in which $m > 0$ and $n < 0$. Let $n = -p$ where $p > 0$. Then

$$a^m a^n = a^m a^{-p} = a^m (a^{-1})^p = (a \ldots a)(a^{-1} \ldots a^{-1})$$

where in the first bracket we have m a's and in the second bracket we have p a^{-1}'s. But

$$(a \ldots a)(a^{-1} \ldots a^{-1}) = (a \ldots a)(aa^{-1})(a^{-1} \ldots a^{-1})$$

$$= (a \ldots a)e(a^{-1} \ldots a^{-1})$$

$$= (a \ldots a)(a^{-1} \ldots a^{-1})$$

where on the right-hand side we have in the first bracket $m - 1$ a's and in the second bracket $p - 1$ a^{-1}'s. Continuing in this way to replace aa^{-1} by e we eventually have

$$a^m a^n = \begin{cases} a \ldots a & m - p \text{ } a\text{'s} & (p < m) \\ e & & (m = p) \\ a^{-1} \ldots a^{-1} & p - m \text{ } a^{-1}\text{'s} & (m < p). \end{cases}$$

Thus for $p < m$ we have

$$a^m a^n = a^{m-p} = a^{m+n},$$

for $m = p$ we have

$$a^m a^n = e = a^0 = a^{m-p} = a^{m+n},$$

and for $m < p$ we have

$$a^m a^n = (a^{-1})^{p-m} = a^{-(p-m)}$$
$$= a^{-p+m} = a^{n+m}.$$

□

Example 9

Let $x, y, z \in G$ and suppose we are required to simplify

$$g = (yx)^{-1}(yx^{-1})^2(xyz)(x^{-3}y^2z)^{-1}.$$

we have

$$g = (x^{-1}y^{-1})(yx^{-1}yx^{-1})(xyz)(x^3y^2z)^{-1} = x^{-1}(y^{-1}y)x^{-1}y(x^{-1}x)y(zz^{-1})y^{-2}x^3$$
$$= x^{-1}x^{-1}yyy^{-2}x^3 = x^{-2}y^2y^{-2}x^3 = x^{-2}x^3 = x.$$

Definition 6

Two elements $a, b \in G$ are said to **commute** if $ab = ba$. If any two elements of G commute then G is said to be **commutative** or **Abelian,** otherwise G is said to be **non-commutative** or **non-Abelian.**

If the operation in a group is written as multiplication then the group may or may not be Abelian. If however an additive notation, with $+$, is employed then it is usual to presume that the operation is commutative.

Example 10

We prove that if $a^2 = e$ for all $a \in G$ then G is necessarily Abelian. By assumption for all $a, b \in G$

$$(ab)^2 = e.$$

Thus

$$ab = (ab)^{-1} = b^{-1}a^{-1} = ba.$$

We now give several further examples of groups.

Examples 11

1. Let $\mathbb{O}(2)$ consist of all matrices of the form T_θ where

$$T_\theta = \begin{pmatrix} \cos\theta & \sin\theta \\ -\sin\theta & \cos\theta \end{pmatrix} \quad (\theta \in \mathbb{R}).$$

We claim that under matrix multiplication $\mathbb{O}(2)$ is a group. We require some elementary trigonometry.

We have

$$
\begin{aligned}
T_\theta T_\varphi &= \begin{pmatrix} \cos\theta & \sin\theta \\ -\sin\theta & \cos\theta \end{pmatrix}\begin{pmatrix} \cos\varphi & \sin\varphi \\ -\sin\varphi & \cos\varphi \end{pmatrix} \\
&= \begin{pmatrix} \cos\theta\cos\varphi - \sin\theta\sin\varphi & \cos\theta\sin\varphi + \sin\theta\cos\varphi \\ -\sin\theta\cos\varphi - \cos\theta\sin\varphi & -\sin\theta\sin\varphi + \cos\theta\cos\varphi \end{pmatrix} \\
&= \begin{pmatrix} \cos(\theta+\varphi) & \sin(\theta+\varphi) \\ -\sin(\theta+\varphi) & -\cos(\theta+\varphi) \end{pmatrix}.
\end{aligned}
$$

This establishes closure and associativity is immediate.

$$T_0 = \begin{pmatrix} \cos 0 & \sin 0 \\ -\sin 0 & \cos 0 \end{pmatrix} = \begin{pmatrix} 1 & 0 \\ 0 & 1 \end{pmatrix} \text{ is the identity.}$$

Since

$$T_{-\theta}T_\theta = T_{-\theta+\theta} = T_0 = \begin{pmatrix} 1 & 0 \\ 0 & 1 \end{pmatrix},$$

$$(T_\theta)^{-1} = T_{-\theta}.$$

Thus $\mathbb{O}(2)$ is a group which is infinite and Abelian since

$$T_\theta T_\varphi = T_{\theta+\varphi} = T_{\varphi+\theta} = T_\varphi T_\theta.$$

2. A mapping $f_{a,b}$ of $\mathbb{R}$ into $\mathbb{R}$ is defined for $a, b \in \mathbb{R}$ by

$$f_{a,b} : x \to a + bx \quad (x \in \mathbb{R}).$$

Then, as in an Exercises 4.1, no. 2

$$f_{c,d} \circ f_{a,b} = f_{c+ad,\,bd} \quad (a, b, c, d \in \mathbb{R}).$$

Let T be the set of such mappings $f_{a,b}$ for which $b \neq 0$. We claim that T is a group. Certainly we have closure since if $b \neq 0$, $d \neq 0$ then $bd \neq 0$ and so

$$f_{c,d} \circ f_{a,b} = f_{c+ad,\,bd} \in T.$$

The circle composition of mappings is associative and $f_{0,1}$ is such that

$$f_{0,1} \circ f_{a,b} = f_{a,b}.$$

For given $a, b \in \mathbb{R}$, $b \neq 0$ we require to find $c, d \in \mathbb{R}$, $d \neq 0$ such that

$$f_{c,d} \circ f_{a,b} = f_{0,1}.$$

But this is only possible if

$$c + ad = 0, \quad bd = 1.$$

However we do have a unique solution, namely

$$d = \frac{1}{b}, \quad c = -\frac{a}{b}.$$

We therefore conclude that T is a group as claimed.

3. Let $X = \mathbb{R} \setminus \{0, 1\}$. Six mappings, e, a, b, c, d, f, are defined on X as follows.

$$e(x) = x, \quad a(x) = \frac{x-1}{x}, \quad b(x) = \frac{1}{1-x} \quad c(x) = 1-x,$$

$$d(x) = \frac{x}{x-1}, \quad f(x) = \frac{1}{x}.$$

It is easy to verify that these mappings are defined on X and fairly easy to verify that they map X into X. We claim that these six mappings under the circle composition of mappings form a group of order 6. Closure is not obvious and must be verified somewhat tediously since there are 36 possible products. Associativity, on the other hand, is clear since the circle composition of mappings is associative. e must play the role of identity. The easiest way to verify closure etc. is to construct a Cayley table. By way of illustration we shall evaluate three products, leaving the reader to contemplate and perhaps to evaluate some at least of the remaining 33.

$$(a \circ f)(x) = a(f(x)) = a\left(\frac{1}{x}\right) = \frac{\frac{1}{x} - 1}{\frac{1}{x}} = \frac{1-x}{1} = 1 - x = c(x),$$

$$(f \circ a)(x) = f(a(x)) = f\left(\frac{x-1}{x}\right) = \frac{x}{x-1} = d(x),$$

$$(b \circ d)(x) = b(d(x)) = b\left(\frac{x}{x-1}\right) = \frac{1}{1 - \frac{x}{x-1}} = \frac{x-1}{x-1-x} = 1-x$$

$$= c(x).$$

Then $a \circ f = c$, $f \circ a = d$, $b \circ d = c$. Eventually we obtain the Cayley table.

	e	a	b	c	d	f
e	e	a	b	c	d	f
a	a	b	e	d	f	c
b	b	e	a	f	c	d
c	c	f	d	e	b	a
d	d	c	f	a	e	b
f	f	d	c	b	a	e

We shall later have occasion to refer to this table. In any subsequent discussion we shall omit $\circ$ and write simply, for example, $af = c$, $fa = d$, $bd = c$, etc.

4. Let $X = \{(x, f) \,|\, x, t \in \mathbb{R}\}$. For each $v \in \mathbb{R}$ we define a mapping T_v where $T_v : X \to X$ by

$$T_v(x, t) = (x - vt, t).$$

We claim that the set S of such mappings is a group. For $u, v \in \mathbb{R}$ we have

$$
\begin{aligned}
(T_u \circ T_v)(x, t) &= T_u(T_v(x, t)) \\
&= T_u(x - vt, t) \\
&= (x - vt - ut, t) \\
&= (x - (u + v)t, t) \\
&= T_{u+v}(x, t).
\end{aligned}
$$

Thus

$$T_u \circ T_v = T_{u+v}$$

from which we obtain closure. The associativity of mappings is assured and T_0 is evidently the identity. Further

$$T_{-v} \circ T_v = T_{-v+v} = T_0.$$

and so

$$T_{-v} = (T_v)^{-1}.$$

S is a group and is Abelian since

$$T_u \circ T_v = T_{u+v} = T_{v+u} = T_v \circ T_u.$$

We remark that this example has its origins in the physical world. If we regard (x, t) as the coordinates of space and time and v as the speed then S is the group of Galilean transformations of classical mechanics (after G. Galilei, 1564–1642). (See also Exercises 4.2, no. 10.)

Definition 7

The group G is said to be **cyclic** if every element of G is a power of some given element of G. This given element is said to **generate**, or to be a **generator** of, the group G.

Thus if G is cyclic we may write
$$G = \{a^n : n = 0, 1, \ldots\}$$
for some $a \in G$. A cyclic group is necessarily Abelian.

Example 12

The groups of orders 1, 2, 3, 4, with Cayley tables as given above, are all cyclic with the exception of the Klein four-group. In the case of the group of order 3 we may put $b = a^2$ and in the case of the cyclic group of order 4 we may put $b = a^3$, $c = a^2$.

We now come to a usage of the word 'order' which differs from its previous usage but which, as we shall later see, is related to that previous usage.

Definition 8

Let $a \in G$. Then a is said to have **finite order** if for some $n \in \mathbb{N}$, $a^n = e$. If $n = 1$ then $a = e$ and e is said to have **order** 1. If for $0 < m < n$, $a^m \neq e$ then a is said to have **order** n. If a does not have finite order then a is said to have **infinite order**.

In a finite group all elements have finite order but if all elements of a group have finite order it does not necessarily follow that the group is finite.

Examples 13

1. The additive group $\mathbb{Z}$ has one element, namely 0, of order 1 and all other elements are of infinite order.
2. The multiplicative group $\mathbb{Q} \setminus \{0\}$ has one element of order 1, namely 1, and one element of order 2, namely -1. All other elements have infinite order. The assertions are true of the group $\mathbb{R} \setminus \{0\}$.
3. The multiplicative group $\mathbb{C} \setminus \{0\}$ has no real elements of finite order other than ± 1 but there are complex elements of finite order. We may identify these elements by use of De Moivre's Theorem (after A. De Moivre, 1667–1754).

Let $z \in \mathbb{C} \setminus \{0\}$ and let $z = r(\cos\theta + i\sin\theta)$ $(r, \theta \in \mathbb{R}, \ r > 0)$. Then we have, for $n \in \mathbb{N}$,

$$z^n = r^n(\cos\theta + i\sin\theta)^n = r^n(\cos n\theta + i\sin n\theta).$$

Hence $z^n = 1$ if and only if

$$r^n \cos n\theta = 1, \ r^n \sin n\theta = 0.$$

This holds if and only if

$$r^n = 1, \cos n\theta = 1, \ \sin n\theta = 0,$$

or

$$r = 1 \text{ and } n\theta = 2k\pi \quad (k \in \mathbb{Z}).$$

Thus $z^n = 1$ if and only if

$$z = \cos\frac{2k\pi}{n} + i\sin\frac{2k\pi}{n}.$$

4. Let G be the group of order 6 with Cayley table as in Examples 11 no. 3. Then

$$a^2 = b, \quad a^3 = ab = e, \quad b^2 = a, \quad b^3 = ab = e, \quad c^2 = d^2 = f^2 = e.$$

Thus e has order 1, a and b have order 3, and c, d and f have order 2.

Definition 9

Let $a, b \in G$. Then a is said to be **conjugate** to b if there exists $x \in G$ such that $x^{-1}ax = b$.

We have therefore defined a relation of conjugacy on a group G and we should therefore not be surprised to obtain the next theorem.

Theorem 7

The relation of conjugacy on a group G is an equivalence relation.

Proof

We have to prove reflexivity, symmetry and transitivity for the given relation.

1. Let $a \in G$. Then $e^{-1}ae = a$ and so a is conjugate to a for all $a \in G$.
2. Let $a, b \in G$ and suppose a is conjugate to b. Then there exists $x \in G$ such that $x^{-1}ax = b$. Let $y = x^{-1}$, then $y^{-1} = (x^{-1})^{-1} = x$. We have

$$y^{-1}x^{-1}axy = y^{-1}by$$

and so

$$(xy)^{-1}a(xy) = y^{-1}by$$

But $xy = e$ and hence

$$a = e^{-1}ae = y^{-1}by.$$

Consequently b is conjugate to a.

3. Let $a, b, c \in G$ and suppose a is conjugate to b and b is conjugate to c. Then there exist $x, y \in G$ such that

$$x^{-1}ax = b, \quad y^{-1}ay = c.$$

Let $z = xy$. Then

$$z^{-1}az = (xy)^{-1}a(xy) = y^{-1}x^{-1}axy = y^{-1}by = c.$$

Hence a is conjugate to c. This completes the proof of conjugacy. □

Definition 10

Under the relation of conjugacy in a group G the equivalence classes are called the **conjugacy** classes.

A typical conjugacy class is $\{x^{-1}ax : x \in G\}$ which is the conjugacy class containing a. In an Abelian group G each conjugacy class consists of a single element since $ax = xa$ $(a, x \in G)$ implies $x^{-1}ax = a$.

Example 14

Suppose we try to find the conjugacy classes in our favourite group of six elements with Cayley table as given above. One class is $\{e\}$. The class containing a is $\{e^{-1}ae, a^{-1}aa, b^{-1}ab, c^{-1}ac, d^{-1}ad, f^{-1}af\}$ which is, in fact, $\{a, a, a, b, b, b\} = \{a, b\}$. The class containing $c \notin \{e\} \cup \{a, b\}$ is $\{e^{-1}ce, a^{-1}ca, b^{-1}cb, c^{-1}cc, d^{-1}cd, f^{-1}cf\}$ which is $\{c, d, f, c, f, d\} = \{c, d, f\}$. The group is therefore partitioned into the three conjugacy classes, $\{e\}$, $\{a, b\}$ and $\{c, d, f\}$.

Definition 11

The element a in G is said to be **self-conjugate** or **central** if $x^{-1}ax = a$ for all $x \in G$. The set of **central** elements is called the **centre** of G and is denoted by $Z(G)$.

Notice that $e \in Z(G)$ and that $Z(G) = G$ if and only if G is Abelian.

Exercises 4.2

1. Let $S = \{0\} \cup \mathbb{N}$. Under the operation of addition is S a group?

2. Prove that the identity e of a group G is the only element of G satisfying the equation $x^2 = x$ $(x \in G)$.

3. Let $a, b, c, d \in G$. Simplify

$$(ab^2)^{-1}(c^2 a^{-1})^{-1}(c^2 b^2 d)(ad)^{-2}a, \quad (abc)^{-1}(ab)^2 d(d^{-1}b^{-1})^2 bdc.$$

4. It is asserted that a group G of order 5, $G = \{e, a, b, c, d\}$, has the following Cayley table

	e	a	b	c	d
e	e	a	b	c	d
a	a	c	d	e	b
b	b	d	e	a	c
c	c	b	a	d	e
d	d	e	c	b	a

Is the assertion correct?

5. In forming the Cayley table of the non-Abelian group of order 6 (Examples 11 no. 3), only three of the 36 products were actually evaluated. Verify the following equations (note omission of $\circ$).

$$ab = e, \quad ac = d, \quad bc = f, \quad bf = d,$$
$$ca = f, \quad cc = e, \quad cf = a, \quad db = f,$$
$$fb = c, \quad ff = e.$$

6. Let $S = \left\{ \cos \dfrac{2k\pi}{3} + i \sin \dfrac{2k\pi}{3} \mid k \in \mathbb{Z} \right\}$.

   Prove that S is a multiplicative group of order 3.

7. Prove that the set of complex numbers

$$S = \left\{ \cos \frac{2k\pi}{n} + i \sin \frac{2k\pi}{n} \mid k \in \mathbb{Z} \right\}$$

   is a cyclic group of order n.

8. A mapping $f : [0, 1] \to [0, 1]$, where $[0, 1]$ is the closed interval of analysis $\{x \mid 0 \le x \le 1\}$, is said to be strictly monotonic if the inequalities $0 \le x_1 < x_2 \le 1$ imply that $f(x_1) < f(x_2)$. Under the circle composition of mappings prove that the set of strictly monotonic mappings of $[0, 1]$ into $[0, 1]$ is a group.

9. Prove that the set G of matrices of the form

$$\begin{pmatrix} a & b \\ -b & a \end{pmatrix} \quad (a, b \in \mathbb{R}, \ a^2 + b^2 \neq 0)$$

is a group under matrix multiplication. What is the centre of this group?

10. (Hard) Like an example above, this problem originates in the physical world. Let c be a given strictly positive constant.

(i) Let $u, v \in R$ be such that $|u| < c$, $|v| < c$.
Prove that $|w| < c$ where

$$w = \frac{u + v}{1 + \dfrac{uv}{c^2}}.$$

(ii) Let $X = \{(x, t) \mid x, t \in \mathbb{R}\}$.
For each $v \in \mathbb{R}$ we define a mapping $T_v : X \to X$ by

$$T_v(x, t) = \left(\frac{x - vt}{\sqrt{1 - \dfrac{v^2}{c^2}}}, \frac{t - \dfrac{v}{c^2} x}{\sqrt{1 - \dfrac{v^2}{c^2}}} \right).$$

Prove that

$$(T_u \circ T_v)(x, t) = \left(\frac{x - wt}{\sqrt{1 - \dfrac{w^2}{c^2}}}, \frac{t - \dfrac{w}{c^2} x}{\sqrt{1 - \dfrac{w^2}{c^2}}} \right).$$

Deduce that the set S of such mappings is a group. This group is in fact the Lorentz group of transformations of relativistic mechanics (after H.A. Lorentz, 1853–1928).

4.3 Subgroups

We have explored some immediate consequences of the definition of the group G. We now wish to look more closely at the structure of G and we begin by examining various sub-structures.

Definition 12

Let H be a non-empty subset of the group G which is also a group under the multiplication in G. Then H is called a **subgroup** of G.

As for subrings, this definition is too descriptive to be useful. We give a convenient criterion.

Theorem 8 Subgroup Criterion

A non-empty subset H of the group G is a subgroup of G if and only if for all $a, b \in H$, $ab \in H$ and $a^{-1} \in H$.

Proof

Suppose H is a subgroup of G. Then H is closed under the multiplication in G and so for all $a, b \in H$, $ab \in H$. Now H has an identity f, say, and then $f^2 = f = fe$ and so $e = f \in H$. Let $a \in H$. Then a has an inverse a', say, in H such that $a'a = e$. But a^{-1} exists in G such that $a^{-1}a = e$. Thus $a^{-1} = a' \in H$.

Conversely if for all $a, b \in H$, $ab \in H$ and $a^{-1} \in H$ then H is evidently closed under the multiplication in G. This multiplication is associative in G and so in H. Let $a \in H$. Then $a^{-1} \in H$ and so $e = aa^{-1} \in H$. For $a \in H$ we have $a^{-1} \in H$ and so, finally, H is a subgroup.   □

Two obvious subgroups of G are G itself and $\{e\}$.

Definition 13

The subgroup H of the group G is said to be a **proper** subgroup if $\{e\} \neq H$ and $H \neq G$.

Examples 15

1. Let $a \in G$. Then $H = \{a^n \mid n \in \mathbb{Z}\}$ is a subgroup of G. To see this observe that typical elements of H are a^m and a^n. Then

$$a^m a^n = a^{m+n} \text{ and } (a^m)^{-1} = a^{-m}$$

and so H is a subgroup of G.

2. Under the usual multiplication, let G be the group $\mathbb{Q} \setminus \{0\}$ and let H be the subset of G given as $\{u^2 \mid u \in \mathbb{Q} \setminus \{0\}\}$.

Then H is certainly a non-empty subset of G. Furthermore if $a, b \in H$ we have $a = x^2$, $b = y^2$ for some $x, y \in \mathbb{Q} \setminus \{0\}$. Then

$$ab = x^2 y^2 = (xy)^2 \text{ and } a^{-1} = (x^2)^{-1} = x^{-2} = (x^{-1})^2$$

Since $(xy)^2 \in H$ and $(x^{-1})^2 \in H$ we may conclude that H is a subgroup.

3. The cyclic group G of order 4, $G = \{e, a, b, c\}$ with Cayley table

	e	a	b	c
e	e	a	b	c
a	a	c	e	b
b	b	e	c	a
c	c	b	a	e

has one proper subgroup, namely $\{e, c\}$.

4. Under matrix multiplication the set G of matrices of the form

$$\begin{pmatrix} a & b \\ -b & a \end{pmatrix} \quad (a, b \in \mathbb{R}, a^2 + b^2 \neq 0)$$

forms a group. The four matrices

$$\begin{pmatrix} 1 & 0 \\ 0 & 1 \end{pmatrix}, \begin{pmatrix} -1 & 0 \\ 0 & -1 \end{pmatrix}, \begin{pmatrix} 0 & 1 \\ -1 & 0 \end{pmatrix}, \begin{pmatrix} 0 & -1 \\ 1 & 0 \end{pmatrix}$$

form a subgroup of G.

Theorem 9

Let H and K be subgroups of the group G. Then $H \cap K$ is a subgroup of G.

Proof

We should observe first of all that $H \cap K$ is indeed non-empty since $e \in H$ and $e \in K$ implies that $e \in H \cap K$. Let $a, b \in H \cap K$. Then $a, b \in H$ and so, as H is a subgroup, $ab \in H$ and $a^{-1} \in H$. Similarly $ab \in K$ and $a^{-1} \in K$. Thus $ab \in H \cap K$ and $a^{-1} \in H \cap K$ from which we conclude that $H \cap K$ is a subgroup. $\square$

A simple adaptation of the proof above shows that the intersection of any collection of subgroups is again a subgroup. However, we do not have what might be thought to be the analogous result for the union of two subgroups; we may note the following result in passing.

Theorem 10

Let H and K be subgroups of the group G. Suppose $H \cup K$ is a subgroup of G. Then either $H \subseteq K$ or $K \subseteq H$.

Proof

For the sake of argument suppose that H is not a subset of K. Then there exists $h \in H$, $h \notin K$. We shall prove that $K \subseteq H$.

Let $a \in K$. Then $a, h \in H \cup K$ and, as $H \cup K$ is a subgroup, $ah \in H \cup K$. We then have two possible alternatives, either $ah \in H$ or $ah \in K$. If $ah \in K$ and since $a^{-1} \in K$ we would have $h = a^{-1}(ah) \in K$ which is contrary to our choice of h. Thus $ah \in H$ and since $h \in H$ we have $h^{-1} \in H$ and so $a = (ah)h^{-1} \in H$. Thus $K \subseteq H$. □

Example 16

Let $D_4 = \{e, a, b, c, d, f, g, h\}$ be the group of order 8 with Cayley table below. The group is called the **dihedral group** of order 8.

	e	a	b	c	d	f	g	h
e	e	a	b	c	d	f	g	h
a	a	b	c	e	f	g	h	d
b	b	c	e	a	g	h	d	f
c	c	e	a	b	h	d	f	g
d	d	h	g	f	e	c	b	a
f	f	d	h	g	a	e	c	b
g	g	f	d	h	b	a	e	c
h	h	g	f	d	c	b	a	e

G and $\{e\}$ are, of course, subgroups. The remaining subgroups are $\{e, b\}$, $\{e, d\}$, $\{e, g\}$, $\{e, f\}$, $\{e, h\}$, $\{e, a, b, c\}$, $\{e, b, d, g\}$ and $\{e, b, f, h\}$; of these subgroups we may verify that the first six are cyclic and the last two are, structurally, Klein four-groups.

We now come to a particular and important type of subgroup.

Lemma 3

Let a be an element of the group G. Let the subset $\{x \in G \,|\, x^{-1}ax = a\}$ be denoted by $C_G(a)$. Then $C_G(a)$ is a subgroup of G containing a.

Proof

We observe first that $C_G(a)$ is non-empty since $a^{-1}aa = a$ and so $a \in C_G(a)$. Let $x, y \in C_G(a)$. Then $x^{-1}ax = a$ and $y^{-1}ay = a$. We have

$$(xy)^{-1}a(xy) = y^{-1}x^{-1}axy = y^{-1}ay = a \text{ and so } xy \in C_G(a).$$

Also from $x^{-1}ax = a$ we have

$$a = xx^{-1}axx^{-1} = xax^{-1} = (x^{-1})^{-1}a(x^{-1}), \text{ from which } x^{-1} \in C_G(a).$$

Thus $C_G(a)$ is a subgroup. $\qquad\qquad\qquad\qquad\qquad\qquad\qquad\qquad\qquad\square$

Definition 14

Let a be an element of the group G. Then $C_G(a) = \{x \in G \,|\, x^{-1}ax = a\}$ is a subgroup called the **centralizer** of a in G.

Since $x^{-1}ax = a$ if and only if $ax = xa$, $C_G(a)$ is the subset of G consisting of all elements 'commuting' with a.

Theorem 11

The centre $Z(G)$ of the group G is a subgroup of G.

Proof

We may follow the argument of Lemma 3. We have

$$Z(G) = \{x \in G \,|\, x^{-1}ax = a \text{ for all } a \in G\}.$$

Thus if $x, y \in Z(G)$ then $x, y \in C_G(a)$ for all $a \in G$ and so $xy, x^{-1}C_G(a)$ for all $a \in G$. Thus $xy, x^{-1} \in Z(G)$ and so $Z(G)$ is a subgroup. $\qquad\qquad\square$

Definition 15

Let X and Y be non-empty subsets of the group G. Let XY be the subset given by

$$XY = \{xy \,|\, x \in X, y \in Y\}.$$

Similarly XYZ, written without brackets as multiplication is associative, denotes the subset given by

$$XYZ = \{xyz \mid x \in X, y \in Y, z \in Z\}.$$

If, in particular, X and Z are each subsets of a single element, $X = \{x\}$ and $Z = \{z\}$, say, then xY is to be written for $\{x\}Y$ and xYz is to be written for $\{x\}Y\{z\}$. Thus

$$xY = \{xy \mid y \in Y\}, \quad Yz = \{yz \mid y \in Y\}.$$

We note that if A, B, X, Y are subsets of G such that $A \subseteq B$ then evidently $XAY \subseteq XBY$. We remark that for a subgroup H of the group G we have necessarily

$$hH = Hh = H \text{ for all } h \in H.$$

Lemma 4

Let H be a subgroup of the group G and let $x \in G$. Then the subset $x^{-1}Hx$ given by $\{x^{-1}hx \mid h \in H\}$ is a subgroup of G.

Proof

Let $a, b \in x^{-1}Hx$. Then there exist $h, k \in H$ such that

$$a = x^{-1}hx, \quad b = x^{-1}kx.$$

But

$$ab = (x^{-1}hx)(x^{-1}kx) = x^{-1}hxx^{-1}kx = x^{-1}hkx,$$

and

$$a^{-1} = x^{-1}h^{-1}x$$

since

$$(x^{-1}h^{-1}x)(a) = (x^{-1}h^{-1}x)(x^{-1}hx) = x^{-1}h^{-1}xx^{-1}hx = x^{-1}h^{-1}hx = x^{-1}x = e.$$

But H is a subgroup and so $hk, h^{-1} \in H$. Thus $ab, a^{-1} \in x^{-1}Hx$ which is therefore a subgroup. $\qquad\square$

Definition 16

Let H be a subgroup of the group G. Any subgroup of the form $x^{-1}Hx$ for some $x \in G$ is called a **conjugate** of H. A subgroup H which coincides with all its conjugates is said to be **normal** in G.

Thus H is normal in G if and only if $x^{-1}Hx = H$ for all $x \in G$ or, equivalently, $xH = Hx$ for all $x \in G$.

A useful characterization of normality is given in the next result.

Theorem 12

Let H be a subgroup of the group G. Then H is normal in G if and only if $x^{-1}Hx \subseteq H$ for all $x \in G$.

Proof

Certainly if H is normal in G the condition holds by definition.

Suppose now $x^{-1}Hx \subseteq H$ for all $x \in G$. We want to show that $H \subseteq x^{-1}Hx$ for all $x \in G$. Now $x^{-1}Hx \subseteq H$ implies that

$$x(x^{-1}Hx)x^{-1} \subseteq xHx^{-1}$$

and so

$$H = (xx^{-1})H(xx^{-1}) \subseteq xHx^{-1}.$$

This proves the result. □

Normal subgroups are usually easier to determine than subgroups which are not normal. The centre of a group is always normal and all subgroups of an Abelian group are normal.

Example 17

In the dihedral group D_4 of order 8 in Example 16, the centre $Z(D_4)$ is $\{e, b\}$. To see this we observe that

$$b^{-1}ab = bab = cb = a,$$

$$b^{-1}cb = bcb = ab = c,$$

$$b^{-1}db = bdb = gb = d, \text{ etc.}$$

The remaining subgroups of order 2 are not normal since, for example,

$$a^{-1}\{e, d\}a = \{a^{-1}ea, a^{-1}da\} = \{e, g\}.$$

The subgroups of order 4 are all normal since we have

$$a^{-1}\{e, b, d, g\}a = \{a^{-1}ea, a^{-1}ba, a^{-1}da, a^{-1}ga\} = \{e, b, g, b\} \text{ etc.}$$

We have proved that the subgroups $\{e, d\}$ and $\{e, g\}$ are conjugate and we may prove that $\{e, f\}$ and $\{e, h\}$ are conjugate.

Any two distinct four-element subgroups have $\{e, b\}$ as their intersection. We note, for future reference, that

$$\{e, d\}\{e, g\} = \{e, b, d, g\} = \{e, g\}\{e, d\},$$

but

$$\{e, d\}\{e, f\} = \{e, c, d, f\} \neq \{e, a, d, f\} = \{e, f\}\{e, d\}.$$

Theorem 13

Let H and K be normal subgroups of the group G. Then $H \cap K$ is a normal subgroup of G.

Proof

We have already shown that $H \cap K$ is a subgroup of G. We merely have to establish the normality of $H \cap K$ in G. Let $x \in G$. Then from $H \cap K \subseteq H$ we have $x^{-1}(H \cap K)x \subseteq H$. Similarly $x^{-1}(H \cap K)x \subseteq K$. Thus $x^{-1}(H \cap K)x \subseteq H \cap K$ and so $H \cap K$ is normal in G. □

Lemma 5

Let H be a subgroup of the group G. Let $N_G(H)$ denote the subset of G given by $\{x \in G \mid x^{-1}Hx = H\}$. Then $N_G(H)$ is a subgroup of G containing H.

Proof

Let $x, y \in N_G(H)$. Then $x^{-1}Hx = H$ and $y^{-1}Hy = H$. Hence

$$(xy)^{-1}H(xy) = y^{-1}x^{-1}Hxy = y^{-1}Hy = H$$

and so $xy \in N_G(H)$. Also

$$H = eHe = (x^{-1})^{-1}x^{-1}Hxx^{-1} = (x^{-1})^{-1}H(x^{-1})$$

and so $x^{-1} \in N_G(H)$. Thus $N_G(H)$ is a subgroup. □

Definition 17

Let H be a subgroup of the group G. Then $N_G(H) = \{x \in G \mid x^{-1}Hx = H\}$ is a subgroup called the **normalizer** of H in G.

$N_G(H)$ is the largest subgroup of G in which H is a normal subgroup. $N_G(H) = H$ if and only if H is a normal subgroup of G.

Theorem 14

Let H and N be subgroups of G and let N be a normal subgroup of G. Then $HN = NH$ and HN is a subgroup of G.

Proof

First we prove that $HN = NH$. We observe that HN consists of all elements of the form hn ($h \in H, n \in N$) and moreover that each such element belongs to NH since $hn = (hnh^{-1})h \in NH$ as N is normal in G, thus $HN \subseteq NH$. Similarly $NH \subseteq HN$ and so $NH = HN$.

Suppose now $a, b \in HN$. Then there exist $h_1, h_2 \in H$ and $n_1, n_2 \in N$ such that $a = h_1 n_1$ and $b = h_2 n_2$. Then

$$ab = h_1 n_1 h_2 n_2 = h_1 (n_1 h_2 n_1^{-1}) n_1 n_2 \in HN$$

since N is a normal subgroup and H is a subgroup and

$$a^{-1} = (h_1 n_1)^{-1} = n_1^{-1} h_1^{-1} \in NH = HN.$$

Thus HN is a subgroup of G. □

Examples 18

1. In the dihedral group $G = D_4$ of order 8 as above, suppose we wish to find $N_G(\{e, d\})$. Reading from the Cayley table, we have $e^{-1}de = d$, $a^{-1}da = g$, $b^{-1}db = d$, $c^{-1}dc = g$, $d^{-1}dd = d$, $f^{-1}df = g$, $g^{-1}dg = d$, $h^{-1}dh = g$. Thus

$$N_G(\{e, d\}) = \{e, b, d, g\}.$$

Alternatively we may observe that

$$N_G(\{e, d\}) \supseteq Z(D_4) = \{e, b\}$$

and so

$$N_G(\{e, d\}) \supseteq \{e, d\}\{e, b\} = \{e, b, d, g\}.$$

But

$$N_G(\{e, d\}) \neq D_4$$

as $\{e, d\}$ is not normal and we already know all subgroups of D_4 so that

$$N_G(\{e, d\}) = \{e, b, d, g\}.$$

Similarly we have

$$N_G(\{e,g\}) = \{e,b,d,g\}, \quad N_G(\{e,f\}) = \{e,b,f,h\}.$$

2. We recall that a square matrix A over $\mathbb{R}$ is **non-singular** if A^{-1} exists, or equivalently if A has non-zero determinant. In the case of a 2×2 matrix

$$A = \begin{pmatrix} a & b \\ c & d \end{pmatrix},$$

A is non-singular if and only if $ad - bc \neq 0$ and then

$$a^{-1} = \frac{1}{ad - bc} \begin{pmatrix} d & -b \\ -c & a \end{pmatrix}.$$

The set of such non-singular 2×2 matrices over $\mathbb{R}$ forms a group under multiplication which is called the **general linear group** of 2×2 matrices over $\mathbb{R}$ and which is denoted by $GL(2, \mathbb{R})$ (for the proof that $GL(2, \mathbb{R})$ is a group see Exercises 4.3, no. 6).

Suppose we want to find the centre $Z(GL(2, \mathbb{R}))$. Then we are looking for all matrices

$$\begin{pmatrix} x & y \\ z & t \end{pmatrix} \quad (xt - yz \neq 0)$$

such that

$$\begin{pmatrix} x & y \\ z & t \end{pmatrix} \begin{pmatrix} a & b \\ c & d \end{pmatrix} = \begin{pmatrix} a & b \\ c & d \end{pmatrix} \begin{pmatrix} x & y \\ z & t \end{pmatrix}$$

for all $a, b, c, d \in \mathbb{R}$, $ad - bc \neq 0$.

Instead of considering arbitrary a, b, c, d we notice that the above has certainly to hold for particular choices of non-singular matrices such as

$$\begin{pmatrix} 1 & 1 \\ 0 & 1 \end{pmatrix} \quad \text{and} \quad \begin{pmatrix} 1 & 0 \\ 1 & 1 \end{pmatrix}.$$

Now

$$\begin{pmatrix} x & y \\ z & t \end{pmatrix} \begin{pmatrix} 1 & 1 \\ 0 & 1 \end{pmatrix} = \begin{pmatrix} 1 & 1 \\ 0 & 1 \end{pmatrix} \begin{pmatrix} x & y \\ z & t \end{pmatrix}$$

implies

$$\begin{pmatrix} x & x+y \\ z & z+t \end{pmatrix} = \begin{pmatrix} x+z & y+t \\ z & t \end{pmatrix}$$

from which

$$x = x + z, x + y = y + t \text{ and so } z = 0, x = t.$$

Similarly from $\begin{pmatrix} 1 & 0 \\ 1 & 1 \end{pmatrix}$ we obtain $y = 0$, $x = t$. Thus for $\begin{pmatrix} x & y \\ z & t \end{pmatrix}$ to be in the centre we must have

$$\begin{pmatrix} x & y \\ z & t \end{pmatrix} = \begin{pmatrix} x & 0 \\ 0 & x \end{pmatrix}.$$

But, on the other hand, we always have

$$\begin{pmatrix} x & 0 \\ 0 & x \end{pmatrix}\begin{pmatrix} a & b \\ c & d \end{pmatrix} = \begin{pmatrix} a & b \\ c & d \end{pmatrix}\begin{pmatrix} x & 0 \\ 0 & z \end{pmatrix}.$$

Hence the centre $Z(GL(2, \mathbb{R}))$ consists precisely of the so-called 'scalar matrices'

$$\begin{pmatrix} x & 0 \\ 0 & x \end{pmatrix} \text{ where } x \neq 0.$$

We claim that

$$H = \left\{ \begin{pmatrix} 1 & a \\ 0 & 1 \end{pmatrix} \mid a \in \mathbb{R} \right\}$$

is an Abelian subgroup of $GL(2, \mathbb{R})$. (For proof see Exercises 4.3, no. 6). Suppose we want to find $N_G(H)$. Then we require to find those matrices $\begin{pmatrix} x & y \\ z & t \end{pmatrix}$ such that

$$\begin{pmatrix} x & y \\ z & t \end{pmatrix}^{-1}\begin{pmatrix} 1 & a \\ 0 & 1 \end{pmatrix}\begin{pmatrix} x & y \\ z & t \end{pmatrix} \in H$$

for all $a \in \mathbb{R}$. Consequently, we require to find those matrices $\begin{pmatrix} x & y \\ z & t \end{pmatrix}$ for which given $a \in \mathbb{R}$ there exists $b \in \mathbb{R}$ such that

$$\begin{pmatrix} x & y \\ z & t \end{pmatrix}^{-1}\begin{pmatrix} 1 & a \\ 0 & 1 \end{pmatrix}\begin{pmatrix} x & y \\ z & t \end{pmatrix} = \begin{pmatrix} 1 & b \\ 0 & 1 \end{pmatrix}$$

or, equivalently,

$$\begin{pmatrix} 1 & a \\ 0 & 1 \end{pmatrix}\begin{pmatrix} x & y \\ z & t \end{pmatrix} = \begin{pmatrix} x & y \\ z & t \end{pmatrix}\begin{pmatrix} 1 & b \\ 0 & 1 \end{pmatrix}.$$

Thus we must have

$$\begin{pmatrix} x + az & y + at \\ z & t \end{pmatrix} = \begin{pmatrix} x & bx + y \\ z & bz + t \end{pmatrix}$$

from which

$$x + az = x, \quad y + at = bx + y, \quad t = bz + t$$

and so

$$az = bz = 0, \quad at = bx.$$

Since a is arbitrary, $z = 0$. But now,

$$\begin{pmatrix} x & y \\ 0 & t \end{pmatrix}^{-1} \begin{pmatrix} 1 & a \\ 0 & 1 \end{pmatrix} \begin{pmatrix} x & y \\ 0 & t \end{pmatrix} \qquad (xt \neq 0)$$

$$= \frac{1}{xt} \begin{pmatrix} t & -y \\ 0 & x \end{pmatrix} \begin{pmatrix} 1 & a \\ 0 & 1 \end{pmatrix} \begin{pmatrix} x & y \\ 0 & t \end{pmatrix}$$

$$= \frac{1}{xt} \begin{pmatrix} tx & t^2 a \\ 0 & xt \end{pmatrix} = \begin{pmatrix} 1 & \dfrac{ta}{x} \\ 0 & 1 \end{pmatrix}.$$

Thus finally

$$N_G(H) = \left\{ \begin{pmatrix} x & y \\ 0 & t \end{pmatrix} \mid xt \neq 0 \right\}.$$

We have defined the centralizer of an element and the normalizer of a subgroup. We may extend the definitions very slightly as follows.

Definition 18

Let A be a non-empty subset of the group G. The **centralizer** and the **normalizer** of A are defined to be the subgroups $C_G(A)$ and $N_G(A)$ given respectively by

$$C_G(A) = \{x \in G \mid x^{-1}ax = a \text{ for all } a \in A\},$$

$$N_G(A) = \{x \in G \mid x^{-1}Ax = A\}.$$

We have asserted that these subsets are subgroups and the assertion should be confirmed.

We have

$$G_G(A) = \bigcap_{a \in A} C_G(a),$$

and, in particular,

$$Z(G) = C_G(G) = \bigcap_{a \in G} C_G(a).$$

Exercises 4.3

1. Write down the Cayley table of a cyclic group G of order 6. Find the proper subgroups of G.

2. Let $\{A, B, C, \ldots\}$ be a set of subgroups of a group G. Prove that the intersection $A \cap B \cap C \ldots$ is a subgroup of G.

3. Find the subgroups of the non-Abelian group of order 6 in Examples 11 no. 3 and Example 14. Determine which are normal. Determine the conjugates of each subgroup.

4. Let $\{A, B, C, \ldots\}$ be a set of normal subgroups of a group G. Prove that $A \cap B \cap C \cap \ldots$ is a normal subgroup of G.

5. Let H be a subgroup of the group G and let $a \in G$. Prove that $a^{-1}Ha$ is a subgroup of G and that

$$\bigcap_{x \in G} x^{-1}Hx$$

is a normal subgroup of G.

6. Prove that the set $GL(2, R)$ of non-singular 2×2 matrices is a group, and that

$$H = \left\{ \begin{pmatrix} 1 & a \\ 0 & 1 \end{pmatrix} : a \in \mathbb{R} \right\}$$

is an Abelian subgroup.

7. Let A be a non-empty subset of the group G. Prove that $C_G(A)$ is a normal subgroup of $N_G(A)$

8. (Hard) We have shown that if N and H are subgroups of G and N is a normal subgroup then NH is a subgroup of G and $NH = HN$. Without assuming that N is a normal subgroup prove that NH is a subgroup if and only if $NH = HN$.

9. (Hard) Let G be a finite group and let H be a proper subgroup of G and let H have n conjugates $H = H_1, H_2, \ldots, H_n$. Prove that the set $H_1 \cup H_2 \cup \ldots \cup H_n$ contains, at most, $n(|H| - 1) + 1$ elements of G. Deduce that G is not equal to the union $H_1 \cup H_2 \cup \ldots \cup H_n$.

4.4 Lagrange's Theorem, Cosets and Conjugacy

We shall consider two partitions of G. The first will be obtained from 'cosets' of a subgroup of G and the second from the conjugacy of elements of G.

Definition 19

Let H be a subgroup of the group G. Let $a \in G$. Then

$$aH = \{ah | h \in H\}$$

is called the **left coset** of H in G determined by a. Similarly

$$Ha = \{ha \mid h \in H\}$$

is called the **right coset** of H in G determined by a.

We note that $a \in Ha$ and that $aH = Ha$ for all $a \in G$ if and only if H is normal in G. We shall prove a number of results on right cosets; it will be evident that analogous results will also hold for left cosets.

Theorem 15

Let H be a subgroup of the group G. Let $a, b \in G$. Then the following statements are equivalent.

1. $Ha = Hb$.
2. There exists $h \in H$ such that $b = ha$.
3. $b \in Ha$.
4. $ba^{-1} \in H$.

Proof

We prove the equivalence of the statements in a cyclic manner.

Suppose 1 holds. Then from $Ha = Hb$ we have $b \in Hb = Ha$ and so there exists $h \in H$ such that $b = ha$. Thus 2 holds.
Suppose 2 holds. Since $b = ha$ for some $h \in H$ we have, by definition, $b \in Ha$ and 3 holds.
Suppose 3 holds. From $b \in Ha$ we have $b = ha$ for some $h \in H$ and so, evidently, $ba^{-1} = haa^{-1} = h \in H$. Thus 4 holds.
Suppose 4 holds. Then $ba^{-1} \in H$ implies $Hba^{-1} = H$ and so $Hb = Hba^{-1}a = Ha$. This completes the proof.   □

Theorem 16

Let H be a subgroup of the group G. Any two right cosets of H in G are either equal or have empty intersection. Consequently the set of right cosets of H in G is a partition of G.

Proof

Let $a, b \in G$. We have to show that either $Ha = Hb$ or $Ha \cap Hb = \emptyset$. Suppose $Ha \cap Hb \neq \emptyset$, then we prove that $Ha = Hb$. Since $Ha \cap Hb \neq \emptyset$ there exists

$x \in Ha \cap Hb$. Hence there exist $h_1, h_2 \in H$ such that $x = h_1 a = h_2 b$. But then $b = h_2^{-1} h_2 b = h_2^{-1} h_1 a$ where $h_2^{-1} h_1 \in H$ as H is a subgroup. But then by Theorem 15 $Ha = Hb$.

G is certainly the union of the cosets of H in G since each element x (say) of G belongs to the coset Hx in G. But Hx is also the only coset to which x belongs since if we find that $x \in Hy$ then necessarily $Hx = Hy$. Thus we may conclude that the set of right cosets of H in G is a partition of G.   □

Example 19

Consider the dihedral group D_4 of order 8 (Example 16 and 17). $H = \{e, f\}$ is a subgroup. $a \notin \{e, f\}$ and $\{e, f\}a = \{ea, fa\} = \{a, d\}$. Choose $b, b \notin \{e, f\} \cup \{e, f\}a$, then $\{e, f\}b = \{eb, fb\} = \{b, h\}$. Choose $c \notin \{e, f\} \cup \{e, f\}a \cup \{e, f\}b$ and then we obtain $G = \{e, f\} \cup \{e, f\}a \cup \{e, f\}b \cup \{e, f\}c$, which expresses G as a union of four mutually disjoint right cosets of H in G.

Definition 20

Let H be a subgroup of the group G. If G is written as a union of mutually disjoint right (left) cosets of H in G then G is said to be written as a **right (left) coset decomposition** of H in G. If there is an infinite number of cosets in a right (left) coset decomposition then H is said to have **infinite index** in G. If there is a finite number of cosets in a coset decomposition then H is said to have **finite index** equal to this number of cosets; we denote this number by $|G : H|$.

At first glance it might appear that the index could depend on whether we were using right or left cosets. However, if we have a right coset decomposition

$$G = Ha_1 \cup Ha_2 \cup \ldots \cup Ha_n$$

then we also have a left coset decomposition, namely,

$$G = a_1^{-1} H \cup a_2^{-1} H \cup \ldots \cup a_n^{-1} H.$$

(For proof of this observation see Exercises 4.4, no. 2). Commonly we take $a_1 = e$ and so Ha_1 is the subgroup H.

Lemma 6

Let H be a finite subgroup of G. Then the number of distinct elements in any coset of H is precisely the number of elements in H.

Proof

Let $a \in H$. Consider Ha. We have to prove that H and Ha have the same number of elements. We claim that the mapping $f : H \to Ha$ given by

$$f(h) = ha \quad (x \in H).$$

is a bijection. Certainly f is a surjection and $f(h_1) = f(h_2)$ $(h_1, h_2 H)$ implies that $h_1 a = h_2 a$ and so $h_1 = h_2$, giving f as an injection. This completes the proof.   □

We now prove one of the seminal theorems in group theory. This theorem is named after J.-L. Lagrange (1736–1813) who, in effect, proved what we would now regard as a special case of the theorem.

Theorem 17 Lagrange's Theorem

Let H be a subgroup of the finite group G. Then the order of H divides the order of G and, in fact, $|G| = |G : H||H|$.

Proof

The group G may be written as a coset decomposition of the cosets of H in G. Suppose $|G : H| = n$. Then we have, say,

$$G = Ha_1 \cup Ha_2 \cup \ldots \cup Ha_n,$$

where the cosets $Ha_1, Ha_2, \ldots, Ha_n$ are distinct. By Lemma 6 the number of elements in each coset is $|H|$ and there are n such cosets. The number of elements in G is $|G|$ and so

$$|G| = n|H| = |G : H||H|$$

giving the result.   □

Remark

If H and K are subgroups of the finite subgroup G where $H \subseteq K \subseteq G$ then

$$|G : H||H| = |G| = |G : K||K| = |G : K||K : H||H|$$

and so

$$|G : H| = |G : K||K : H|.$$

Examples 20

1. Let the order of G be a prime p. Then G has only the subgroups G and $\{e\}$ since if H is a subgroup of G, $|H|$ divides p and so $|H| = p$ or $|H| = 1$ giving the result.

2. We have listed the subgroups of the dihedral group D_4 of order 8. Note that all subgroups are of orders 1, 2, 4 or 8.

3. It is fairly easy to prove that $\mathbb{Z}$, considered as an additive group, has the subset $6\mathbb{Z} = \{6x \mid x \in \mathbb{Z}\}$ as a proper subgroup. The integer a belongs to the coset $6\mathbb{Z} + a = \{6x + a \mid x \in \mathbb{Z}\}$ and a and b ($b \in \mathbb{Z}$) belong to the same coset if and only if $6\mathbb{Z} + a = 6\mathbb{Z} + b$. Thus a and b belong to the same coset if and only if $a - b \in 6\mathbb{Z}$ or 6 divides $a - b$. The six cosets of $6\mathbb{Z}$ are

$$6\mathbb{Z}, \quad 6\mathbb{Z} + 1, \quad 6\mathbb{Z} + 2, \quad 6\mathbb{Z} + 3, \quad 6\mathbb{Z} + 4, \quad 6\mathbb{Z} + 5.$$

We now prove a result on cyclic groups which depends on the Division Algorithm for integers.

Theorem 18

An element of order n of a group G generates a cyclic subgroup of order n of G, and conversely.

Proof

Let $a \in G$ and suppose a has order n. Then $e, a, a^2, \ldots, a^{n-1}$ are distinct and $a^n = e$. Let $H = \{a^m \mid m \in \mathbb{Z}\}$. Then H is the cyclic subgroup generated by a.

Consider now a typical element a^m of H. By the Division Algorithm for integers there exist $q, r \in \mathbb{Z}$, $0 \le r < n$, such that

$$m = qn + r.$$

Then

$$a^m = a^{qn+r} = a^{qn}a^r = (a^n)^q a^r = e^q a^r = ea^r = a^r.$$

Thus $H = \{e, a, \ldots, a^{n-1}\}$ and so a generates a cyclic subgroup of order n.

Conversely if H is a cyclic subgroup of G of order n then H is generated by an element a of order p (say). But an element a of order p generates a subgroup of order p and so, as H has order n, we must have $p = n$, thereby completing the proof.   □

Theorem 19

A finite cyclic group G has subgroups of all possible orders.

Proof

Suppose G has order n. Then by Lagrange's Theorem any subgroup must have order m where m divides n. We wish to show that a subgroup of order m exists. Let $m = nd$ ($d \in \mathbb{N}$). Let G be generated by a and let H be generated by a^d,

$$H = \{(a^d)^r \mid r \in \mathbb{Z}\}.$$

Now $(a^d)^m = a^{md} = a^n = e$ and $e, a^d, a^{2d}, \ldots, a^{(m-1)d}$ are distinct since a has order $n = md$. Thus a^d has order m and so, by Theorem 18, a^d generates a subgroup, namely H, of order m.   □

We now turn to the second of our partitions on the group G. We recall that if $a, b \in G$ then a is conjugate to b if there exists $x \in G$ such that $x^{-1}ax = b$ and that we proved that conjugacy is an equivalence relation on G (Theorem 7); this equivalence relation yields our partition of G.

Definition 21

Let a be an element of the group G. The set of all elements of G conjugate to a, namely $\{x^{-1}ax \mid x \in G\}$, is called the **conjugacy class** of G containing a.

The group G is therefore the disjoint union of its conjugacy classes. We now give a formula for the size of a finite conjugacy class in G.

Theorem 20

Let a be an element of the group G. Suppose the centralizer of a in $G, C_G(a)$, has index n in G. Then a has precisely n conjugates.

Proof

Let

$$G = C_G(a)x_1 \cup C_G(a)x_2 \cup \ldots \cup C_G(a)x_n$$

be a coset decomposition of $C_G(a)$ in G. We claim that the conjugates of a are precisely

$$x_1^{-1}ax_1, \; x_2^{-1}ax_2, \; \ldots, \; x_n^{-1}ax_n.$$

Certainly if $x \in G$ then x belongs to some coset, say $x = cx_k$ for some $c \in C_G(a)$ and some x_k $(1 \leq k \leq n)$. Then

$$x^{-1}ax = (cx_k)^{-1}a(cx_k) = x_k^{-1}c^{-1}acx_k = x_k^{-1}ax_k.$$

Thus the conjugates of a are

$$x_1^{-1}ax_1, \ x_2^{-1}ax_2, \ \ldots, \ x_n^{-1}ax_n.$$

We now have to show that these conjugates are distinct. Suppose for some i, j

$$x_i^{-1}ax_i = x_j^{-1}ax_j.$$

Then

$$x_j x_i^{-1} a x_i x_j^{-1} = a$$

and so

$$(x_i x_j^{-1})^{-1} a (x_i x_j) = a.$$

But this implies that $x_i x_j^{-1} \in C_G(a)$ and hence x_i and x_j belong to the same coset of $C_G(a)$ in G. This is impossible unless $i = j$. Hence we obtain the result. ☐

Corollary

Let a be an element of the group G. Then the number of elements in the conjugacy class of a divides $|G|$.

Proof

The number is $|G : C_G(a)|$ which, by Lagrange's Theorem, divides $|G|$. ☐

Example 21

Suppose we try to find the conjugacy classes in the dihedral group D_4 of order 8. Since $Z(D_4) = \{e, b\}$, $\{e\}$ and $\{b\}$ are two conjugacy classes. Consider the element a of D_4. We need to find $C_G(a)$. We notice that $\{e, a, b, c\}$ is an Abelian subgroup of G and so $\{e, a, b, c\} \subseteq C_G(a)$. This implies that 4 divides $|C_G(a)|$ which, of course, divides 8. Thus $|C_G(a)| = 4$ or 8. But $|C_G(a)| = 8$ implies that $a \in Z(D_4)$ which is false. Thus $|C_G(a)| = 4$ and so $C_G(a) = \{e, a, b, c\}$. A suitable coset decomposition is

$$G = C_G(a) \cup C_G(a)d$$

and so $\{a, d^{-1}ad\} = \{a, c\}$ is a conjugacy class. The centralizer $C_G(d)$ of the element d is $\{e, b, d, g\}$ and the conjugacy class containing d is $\{d, g\}$. Similarly we may calculate the remaining conjugacy classes and so finally obtain as the conjugacy classes of D_4, $\{e\}$, $\{b\}$, $\{a,b\}$, $\{d,g\}$ and $\{f,h\}$.

Exercises 4.4

1. In the same way that we proved that the number of conjugates of an element of a group G is the index of the centralizer of that element in the group G, prove that if a subgroup H of G has n conjugates in G then the normalizer, $N_G(H)$, has index n in G.

2. If
$$G = Ha_1 \cup Ha_2 \cup \ldots \cup Ha_n$$
is a right coset decomposition of a subgroup H in the group G, prove that
$$a_1^{-1}H \cup a_2^{-1}H \cup \ldots \cup a_n^{-1}H$$
is a left coset decomposition of H in G.

3. What are the cosets of $12\mathbf{Z} = \{12x \mid x \in \mathbf{Z}\}$ considered as a subgroup of the additive group $\mathbf{Z}$?

4.5 Homomorphisms

We are here concerned with mappings which preserve algebraic structure. Such mappings are called 'homomorphisms' (Greek: *homo* same, *morph* form). More precisely we have the following definition which we make for semigroups.

Definition 22

Let S and T be semigroups and let $f : S \to T$ be a mapping such that
$$f(xy) = f(x)f(y)$$
for all $x, y \in S$. Then f is said to be a **homomorphism** from S into T.

Lemma 7

Let S and T be semigroups and let $f : S \to T$ be a homomorphism. Then $f(S)$ is a semigroup (strictly $f(S)$ is a subsemigroup of T).

Proof

Let $a, b \in f(S)$. Then there exist $x, y \in S$ such that $f(x) = a$, $f(y) = b$. Thus

$$ab = f(x)f(y) = f(xy) \in f(S).$$

Thus $f(S)$ is closed under multiplication and, since $f(S)$ is a subset of T, the multiplication is associative. Thus $f(S)$ is a semigroup. $\square$

For the remainder of this chapter we shall be concerned with homomorphisms of groups.

Theorem 21

Let G and H be groups and let $f : G \to H$ be a homomorphism. Then the following hold.

1. If e_G and e_H are the identities of G and H respectively, then $f(e_G) = e_H$.
2. For all $x \in G$, $[f(x)]^{-1} = f(x^{-1})$ where $[f(x)]^{-1}$ is the inverse of $f(x)$ in H and x^{-1} is the inverse of x in G.
3. $f(G)$ is a subgroup of H.
4. Let $K = \{x \in G \mid f(x) = e_H\}$. Then K is a normal subgroup of G.

Proof

We note the necessity of distinguishing the identities in the two groups G and H.

1. Since $f(e_G)f(e_G) = f(e_G e_G) = f(e_G)$, we conclude that $f(e_G) = e_H$.
2. Since $f(x^{-1})f(x) = f(x^{-1}x) = f(e_G) = e_H$ $(x \in G)$, we have $f(x^{-1}) = [f(x)]^{-1}$.
3. By Lemma 7, $f(G)$ is closed under multiplication in H. Let now $a \in f(G)$. Then $a = f(x)$ for some $x \in G$ and $a^{-1} = [f(x)]^{-1} = f(x^{-1}) \in f(G)$. Thus $f(G)$ is a subgroup of H.
4. Since $f(e_G) = e_H$, K is non-empty. Let $x, y \in K$, then $f(x) = f(y) = e_H$. Hence

$$f(xy) = f(x)f(y) = e_H e_H = e_H.$$

and

$$f(x^{-1}) = [f(x)]^{-1} = [e_H]^{-1} = e_H.$$

Consequently K is a subgroup of G. To prove that K is a normal subgroup let $a \in K$, $x \in G$. Then

$$f(x^{-1}ax) = f(x^{-1})f(ax) = f(x^{-1})f(a)f(x)$$
$$= [f(x)]^{-1}e_H f(x) = [f(x)]^{-1}f(x) = e_H,$$

and so $x^{-1}ax \in K$. Hence K is a normal subgroup of G. $\qquad\qquad\square$

Definition 23

Let G and H be groups and let $f : G \to H$ be a homomorphism. Then the subgroup $K = \{x \in G \,|\, f(x) = e_H\}$ is a normal subgroup of G which is called the **kernel** of f, written Ker f.

Examples 22

1. Let G be a group, $G \neq \{e\}$. Then we always have at least two homomorphisms. First we have the mapping f, easily checked to be a homomorphism, given by $f(x) = e$ $(x \in G)$. Second, we have the mapping g defined by $g(x) = x$ $(x \in G)$, again easily verified to be a homomorphism.

2. Let G be the cyclic group of order 4 generated by a, $G = \{e_G, a, a^2, a^3\}$ and $a^4 = e_G$. Let H be the cyclic group of order 2 generated by b, $H = \{e_H, b\}$, $b^2 = e_H$. Then we assert that the mapping $f : G \to H$ given by

$$f(e_G) = e_H, \quad f(a) = b, \quad f(a^2) = e_H, \quad f(a^3) = b$$

is a homomorphism. We require to establish this assertion. For example we have

$$f(aa^3) = f(a^4) = f(e_G) = e_H = bb = f(a)f(a^3).$$

We may similarly verify the other possibilities.

3. Let $\mathbb{C} \setminus \{0\}$ and $\mathbb{R} \setminus \{0\}$ be the multiplicative groups of non-zero complex and real numbers respectively. Then the mapping $f : \mathbb{C} \setminus \{0\} \to \mathbb{R} \setminus \{0\}$ given by

$$f(z) = |z| \quad (z \in C \setminus \{0\})$$

is a homomorphism since, by a property of the modulus (absolute value) of complex numbers, we have

$$f(z_1 z_2) = |z_1 z_2| = |z_1||z_2| = f(z_1)f(z_2) \quad (z_1, z_2 \in \mathbb{C}).$$

4. Consider the non-Abelian group G of order 6 with Cayley table given in Example 6 no. 3. Define $p : G \to \{1, -1\}$ by

$$p(e) = 1, \quad p(a) = 1, \quad p(b) = 1, \quad p(c) = -1, \quad p(d) = -1, \quad p(f) = -1.$$

Then we may verify that p is a homomorphism, for example

$$p(df) = p(b) = 1 = (-1)(-1) = p(d)p(f).$$

5. The set G of matrices of the form

$$\begin{pmatrix} x & y \\ 0 & t \end{pmatrix} \quad (x, y, t \in \mathbb{R},\ xt \neq 0)$$

has been shown to be a group under matrix manipulation. We claim that the mapping $f : G \rightarrow \mathbb{R} \setminus \{0\}$ given by

$$f : \begin{pmatrix} x & t \\ 0 & t \end{pmatrix} \rightarrow xt$$

is a homomorphism. Certainly we have

$$f\left(\begin{pmatrix} x_1 & y_1 \\ 0 & t_1 \end{pmatrix} \begin{pmatrix} x_2 & y_2 \\ 0 & t_2 \end{pmatrix} \right) = f\left(\begin{pmatrix} x_1 x_2 & x_1 y_2 + y_1 t_2 \\ 0 & t_1 t_2 \end{pmatrix} \right)$$

$$= (x_1 x_2)(t_1 t_2) = (x_1 t_1)(x_2 t_2)$$

$$= f\left(\begin{pmatrix} x_1 & y_1 \\ 0 & t_1 \end{pmatrix} \right) f\left(\begin{pmatrix} x_2 & y_2 \\ 0 & t_2 \end{pmatrix} \right)$$

giving the result. (We note that xt is the determinant of $\begin{pmatrix} x & y \\ 0 & t \end{pmatrix}$, see next example).

6. We have defined $GL(2, \mathbb{R})$ as the group of non-singular 2×2 matrices. The mapping $f : GL(2, \mathbb{R}) \rightarrow \mathbb{R}$ given by $f(A) = \det A$, for any non-singular 2×2 matrix A, is a homomorphism since for 'square' matrices A and B, $\det(AB) = \det A \det B$.

7. The existence of certain homomorphisms imposes conditions on the group or groups involved. For example, suppose the mapping $f : x \rightarrow x^{-1}$ $(x \in G)$ is a homomorphism of the group G into itself. Then for $x, y \in G$,

$$x^{-1}y^{-1} = f(x)f(y) = f(xy) = (xy)^{-1} = y^{-1}x^{-1},$$

and so

$$yx = (x^{-1}y^{-1})^{-1} = (y^{-1}x^{-1})^{-1} = xy.$$

Thus if f is a homomorphism G is necessarily Abelian.

Associated to each homomorphism of a group we have a normal subgroup called the kernel. Given a normal subgroup of a group we shall see that we may construct, from the cosets of the normal subgroup, another group onto which the original group may be homomorphically mapped with kernel equal to the normal subgroup. We shall explore for a given group this tight relationship between groups onto which the given group may be homomorphically mapped and normal subgroups of the given group.

Lemma 8

Let G and H be groups and let $f : G \to H$ be a homomorphism with kernel K. Let a and b be elements of G. Then $f(a) = f(b)$ if and only if $Ka = Kb$.

Proof

Suppose $f(a) = f(b)$. Then
$$f(ab^{-1}) = f(a)f(b^{-1}) = f(a)[f(b)]^{-1} = f(b)[f(b)]^{-1} = e_H.$$
Hence $ab^{-1} \in K$ and so $Ka = Kb$. Conversely if $Ka = Kb$ then $b = ka$ for some $h \in K$ and so
$$f(b) = f(ka) = f(k)f(a) = e_H f(a) = f(a). \qquad \square$$

Theorem 22

Let H be a normal subgroup of the group G. Let G/H denote the set of cosets of H in G. A binary operation is defined on G/H as follows. Let Ha and Hb $(a, b \in G)$ be cosets of H in G and define $(Ha)(Hb) = Hab$. Then this binary operation is well-defined and under this operation G/H is a group. Further the mapping $G \to G/H$ given by $a \to Ha$ $(a \in G)$ is a homomorphism of G onto G/H with kernel H.

Proof

It is not immediately obvious that the binary operation on G/H is well-defined. In order that the definition should be well-founded we have to show that if $Ha = Ha'$ and $Hb = Hb'$ $(a, a', b, b' \in G)$ then $Hab = Ha'b'$. Now under the assumption that $Ha = Ha'$ and $Hb = Hb'$ we have that $a' = h_1 a$ and $b' = h_2 b$ for some $h_1, h_2 \in H$ respectively. Then
$$a'b' = h_1 a h_2 b = h_1 a h_2 a^{-1} ab = hab$$
where $h = h_1(ah_2 a^{-1}) \in H$ since H is a normal subgroup of G. Then
$$Ha'b' = Hhab = Hab$$
and so we have vindicated our claim that the definition is well-founded.

We now show that G/H is a group. Certainly, by construction, G/H is closed under the defined multiplicative binary operation. Associativity follows easily since for $a, b, c \in G$.

$$(HaHb)Hc = HabHc = H(ab)c = Ha(bc) = Ha(Hbc) = Ha(HbHc).$$

The coset H plays the role of the identity element of G/H as

$$HHa = HeHa = Hea = Ha \quad (a \in G).$$

The inverse of Ha $(a \in G)$ is given by Ha^{-1} since

$$(Ha^{-1})(Ha) = Ha^{-1}a = He = H.$$

Thus G/H is a group.

The mapping $f : G \to G/H$ given by $f(a) = Ha$ $(a \in G)$ is obviously surjective and is a homomorphism since

$$f(ab) = Hab = HaHb = f(a)f(b) \quad (a, b \in G).$$

Finally

$$\text{Ker } f = \{a \in G \,|\, f(a) = H\}$$
$$= \{a \in G \,|\, Ha = H\} = H. \qquad \square$$

Definition 24

Let H be a normal subgroup of the group G. The group G/H, as constructed above, is said to be a **factor-group** of G.

We note that if H has infinite index in G then G/H is infinite, whereas if H has finite index $|G : H|$ in G then $|G/H| = |G : H|$.

Definition 25

Let G and H be groups and let $f : G \to H$ be a homomorphism. If f is surjective $(f(G) = H)$ then f is said to be an **epimorphism** (Greek: *epi* upon). If f is injective $(f(a) = f(b)$ implies $a = b, a, b \in G)$ then f is said to be a **monomorphism** (Greek: *mono* alone). If f is bijective (surjective and injective) then f is said to be an **isomorphism** (Greek: *iso* equal).

Example 23

The Klein four-group $G = \{e, a, b, c\}$ with Cayley table

	e	a	b	c
e	e	a	b	c
a	a	e	c	b
b	b	c	e	a
c	c	b	a	e

is isomorphic to the group H consisting of the four matrices

$$\begin{pmatrix} 1 & 0 \\ 0 & 1 \end{pmatrix}, \begin{pmatrix} -1 & 0 \\ 0 & -1 \end{pmatrix}, \begin{pmatrix} 0 & -1 \\ 1 & 0 \end{pmatrix}, \begin{pmatrix} 0 & 1 \\ -1 & 0 \end{pmatrix}.$$

The isomorphism is given by the mapping

$$e \rightarrow \begin{pmatrix} 1 & 0 \\ 0 & 1 \end{pmatrix}, \quad a \rightarrow \begin{pmatrix} -1 & 0 \\ 0 & -1 \end{pmatrix}, \quad b \rightarrow \begin{pmatrix} 0 & -1 \\ 1 & 0 \end{pmatrix}, \quad c \rightarrow \begin{pmatrix} 0 & 1 \\ -1 & 0 \end{pmatrix}.$$

Isomorphic groups have the same group-theoretical structure although they may differ in other material aspects.

Example 24

Consider the dihedral group D_4 of order 8, the Cayley table of which we have used frequently. The centre $Z(D_4) = \{e, b\}$ and since the centre is a normal subgroup we may form the group $D_4/Z(D_4)$. We shall see that this group is isomorphic to the Klein four-group.

Let us denote the cosets of $Z(D_4)$ in D_4 by capital letters. Thus let

$$E = Z(D_4) = \{e, b\},$$
$$A = Z(D_4)a = \{e, b\}a = \{a, ba\} = \{a, c\},$$
$$B = Z(D_4)d = \{e, b\}d = \{d, bd\} = d, g\},$$
$$C = Z(D_4)f = \{e, b\}f = \{f, bf\} = \{f, h\}.$$

We may now draw up a Cayley table for $D_4/Z(D_4) = \{E, A, B, C\}$. For example

$$A^2 = Z(D_4)dZ(D_4)d = Z(D_4)d^2 = Z(D_4)e = E,$$
$$AB = Z(D_4)aZ(D_4)d = Z(D_4)ad = Z(D_4)f = C,$$

finally obtaining

	E	A	B	C
E	E	A	B	C
A	A	E	C	B
B	B	C	E	A
C	C	B	A	E

which is the Klein four-group.

We now come to the first of two important results in this section.

Theorem 23 First Isomorphism Theorem

Let G and H be groups and let $f : G \to H$ be an epimorphism with kernel K. Then G/K and H are isomorphic.

Proof

The kernel K of the epimorphism f is a normal subgroup of G and so we also have an epimorphism $g : G \to G/K$. Diagrammatically we have

$$G \xrightarrow{\ f\ } H$$
$$g \downarrow$$
$$G/K \quad ,$$

What we want to do is to construct an isomorphism h from G/K to H. We cannot compose g with f. In a manner of speaking what we do is to reverse backwards along the arrow from G/K to G and go directly along the arrow from G to H. Once we have established that this process does yield a well-defined mapping h it is fairly easy to prove that the h so constructed is an isomorphism.

Thus let Ka $(a \in G)$ be an element of G/K. We would like to define a mapping $h : G/K \to H$ by letting

$$h(Ka) = f(a).$$

This definition only makes sense if $Ka = Kb$ $(a, b \in G)$ implies that $f(a) = f(b)$. But $Ka = Kb$ implies that $b = ka$ where $k \in K$ and then

$$f(b) = f(ka) = f(k)f(a) = e_H f(a) = f(a).$$

Consequently our definition is in fact soundly based.

The mapping h is surjective since if $c \in H$ it follows that, as f is an epimorphism, there exists $a \in G$ such that $f(a) = c$ and then $h(Ka) = f(a) = c$.

The mapping h is injective since if $h(Ka) = h(Kb)$ $(a, b \in G)$ then we have $f(a) = f(b)$ and so $f(ab^{-1}) = f(a)f(b)^{-1} = e_H$ from which $ab^{-1} \in K$ and $Ka = Kb$. It remains now to show that h is a homomorphism. Let $a, b \in G$, then

$$h(KaKb) = h(Kab) = f(ab) = f(a)f(b) = h(Ka)h(Kb).$$

Thus finally we conclude that h is an isomorphism.   □

Examples 25

1. Let G and H be groups and let $f : G \to H$ be a homomorphism. Then $f(G)$ is a subgroup of H and $f(G)$ is isomorphic to G/K.

2. Let G and H be finite groups and let $f : G \to H$ be an epimorphism. Then $|H|$ divides $|G|$ since

$$|G| = |\text{Ker } f||G : \text{Ker } f| = |\text{Ker } f||G/K| = |\text{Ker } f||H|.$$

3. Let H be a subgroup of the centre $Z(G)$ of the group G. Suppose G/H is cyclic. Then there exists $a \in G$ such that $G/H = \{(Ha)^n | n \in \mathbb{Z}\}$. But $(Ha)^n = Ha^n$ and so $G/H = \{Ha^n | n \in \mathbb{Z}\}$. Now this fact implies that G is Abelian since if we have $x, y \in G$ then $x = h_1 a^n$, $y = h_2 a^m$ for some $h_1, h_2 \in H$ and $m, n \in \mathbb{Z}$, and therefore, as $H \subseteq Z(G)$,

$$xy = (h_1 a^n)(h_2 a^m) = h_1 h_2 a^n a^m = h_2 h_1 a^m a^n = h_2 a^m h_1 a^n = yx.$$

We now arrive at the second of our important theorems.

Theorem 24 Second Isomorphism Theorem

Let G be a group, let H and N be subgroups of G and let N be normal in G. Then the factor-groups NH/N and $H/(N \cap H)$ are isomorphic.

Proof

Notice that since N is normal in G, NH is a subgroup and $N \cap H$ is a normal subgroup of H. The following diagram may assist in the understanding of the statement of the theorem:

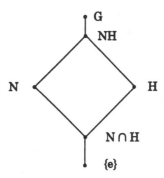

Each subgroup is shown to lie above any subgroup which it contains. Such diagrams in group theory are sometimes called **Hasse diagrams** (after H. Hasse, 1898–1979).

We first define a mapping $p : H \to NH/N$ and show that this mapping p is an epimorphism. For $h \in H$ we define $p(h)$ to be the coset Nh of N in the subgroup $NH = \{nh \mid n \in N, h \in H\}$. Certainly p is surjective. But p is also a homomorphism since

$$p(h_1 h_2) = Nh_1 h_2 = Nh_1 Nh_2 = p(h_1)p(h_2) \ (h_1, h_2 \in H).$$

Thus p is an epimorphism. Let $K = \operatorname{Ker} p$. Then, by the First Isomorphism Theorem, H/K is isomorphic to NH/N. In order to complete the proof we have to prove that $K = N \cap H$. Let $h \in H \cap N$. Then $h \in N$ and so $Nh = N$ and thus $h \in K$. On the other hand, if $k \in K$ then $Nk = N$, and so we have $k \in N$ and finally $k \in N \cap H$. Thus $K = N \cap H$. □

Some groups admit of only two epimorphisms, namely the epimorphism in which all elements of the group are mapped into the number 1 and the isomorphism of the group with itself. Since the existence of epimorphisms is directly related to the existence of normal subgroups we make the following definition.

Definition 26

A non-trivial group which has no proper normal subgroups is said to be **simple**.

Example 26

Suppose we consider a non-trivial group G which is both simple and Abelian. We recall that in an Abelian group all subgroups are normal. Hence our present group G has no proper subgroups. Let $a \in G$, $a \neq e$. Then a generates a non-trivial cyclic subgroup which must be G. If G is infinite then a^2 generates a proper subgroup and we would have a contradiction. Hence G is a finite cyclic group. Let p be a prime divisor of the order of G. Then G has a subgroup of order p which must be G itself. Hence, to summarize, a non-trivial simple Abelian group is cyclic of prime order. For obvious reasons the mathematical interest is in non-Abelian simple groups.

Exercises 4.5

1. Let G, H and K be groups. Let $f : G \to H$ and $g : H \to K$ be given homomorphisms. Prove that $g \circ f$ is a homomorphism.

2. Let G be a group and suppose that the mapping $f : G \to G$, which is given by $f(a) = a^2$ $(a \in G)$, is a homomorphism. Prove that G is Abelian.

3. Let G be a group with centre $Z(G)$. Suppose there exists $a \in G$ such that $x^{-1}a^{-1}xa \in Z(G)$ for all $x \in G$. Prove that the mapping $x \to x^{-1}a^{-1}xa$ is a homomorphism of G into $Z(G)$.

4. Let N and H be subgroups of the group G where $N \subseteq H \subseteq G$ and N is normal in G. Prove that H/N is a subgroup of G and that every subgroup of G/N is of the form H/N for some subgroup H of G. Prove that H/N is a normal subgroup of G/N if and only if H is a normal subgroup of G.

5. The quaternion group Q of order 8 is given by $Q = \{e, a, b, c, d, f, g, h\}$ with Cayley table

	e	a	b	c	d	f	g	h
e	e	a	b	c	d	f	g	h
a	a	e	c	b	f	d	h	g
b	b	c	a	e	g	h	f	d
c	c	b	e	a	h	g	d	f
d	d	f	h	g	a	e	b	c
f	f	d	g	h	e	a	c	b
g	g	h	d	f	c	b	a	e
h	h	g	f	d	b	c	e	a

What is the centre $Z(Q)$ of Q? Find all subgroups of Q and construct a Hasse diagram. Determine the conjugacy classes. Identify the group $Q/Z(Q)$.

5

Rings

We first formulated the abstract concept of a ring from our considerations of the integers and of polynomials. It is our aim to carry forward and to develop the concept of a ring, for which the integers will continue to serve as a useful exemplar.

5.1 Arithmetic Modulo n

Let us, for the present, focus our attention on the prime 3 in the ring of integers $\mathbf{Z}$. Suppose we were to consider only the integers 0, 1, 2. As it stands the set $\{0, 1, 2\}$ is not a ring since, for example, $2 + 2 = 4 \notin \{0, 1, 2\}$. However, suppose we perform addition and multiplication 'modularly' with regard to the prime 3, that is to say we replace any sum or product of the numbers 0, 1, 2 by the corresponding remainder on division by 3. Thus we would have

$$2 + 2 \equiv 1 (\text{mod } 3)$$

since $2 + 2 = 4 = 3 + 1$, where to emphasise that we are performing 'addition' and 'multiplication' modularly we use the symbol $\equiv$ which denotes 'congruence' and we also write 'mod 3'. We may compile modular addition and multiplication tables as follows.

Sum	0	1	2
0	0	1	2
1	1	2	0
2	2	0	1

Product	0	1	2
0	0	0	0
1	0	1	2
2	0	2	1

As we shall shortly see $\{0, 1, 2\}$ with these tables becomes a ring. We now require to make rather more precise some of the notions above.

Definition 1

Let n be a given non-zero integer. Let a and b be two integers. We say that a is **congruent** to b if n divides $a - b$. We write

$$a \equiv b \pmod{n}.$$

Otherwise we say that a is **not congruent** to b modulo n if n does not divide $a - b$.

Example 1

Since 24 divides $48 = 83 - 35$ we have

$$83 \equiv 35 \pmod{24}.$$

Since 17 does not divide $96 = 122 - 26$ it is evident that

$$122 \equiv 26 \pmod{17} \text{ is false.}$$

The next lemma summarises some obvious facts.

Lemma 1

Let n be a given non-zero integer. Then the following statements hold.

1. For all integers $a, a \equiv a \pmod{n}$.
2. Let a and b be integers such that $a \equiv b \pmod{n}$. Then $b \equiv a \pmod{n}$.
3. Let a, b and c be integers such that $a \equiv b \pmod{n}$ and $b \equiv c \pmod{n}$. Then $a \equiv c \pmod{n}$.
4. The relation of congruence is an equivalence relation on $\mathbb{Z}$.

Proof

Certainly n divides $0 = a - a$ for all integers a and so we obtain 1. If for integers a and b, n divides $a - b$ then n divides $b - a$, and 2 follows. To prove 3 we remark that if n divides $a - b$ and n divides $b - c$ then n divides $a - c = (a - b) + (b - c)$ giving 3. Statement 4 is an immediate consequence of the three preceding statements.   □

By Lemma 1, for a given non-zero integer n, the set of integers is partitioned into equivalence classes of integers. If $a \in \mathbb{Z}$ the equivalence class containing a is

$$\{x \in \mathbb{Z} \,|\, x \equiv a \;(\mathrm{mod}\; n)\} = \{a + nk \,|\, k \in \mathbb{Z}\}.$$

Notation

For a given non-zero integer n the equivalence class of integers containing an integer a is denoted by $\bar{a}$.

$$\bar{a} = \{a + kn \,|\, k \in \mathbb{Z}\}.$$

Example 2

Let $n = 12$. Then

$$\bar{4} = \{4 + 12k \,|\, k \in \mathbb{Z}\} = \{\ldots - 20, -8, 4, 16, 28, \ldots\},$$
$$\overline{27} = \{27 + 12k \,|\, k \in \mathbb{Z}\} = \{\ldots, 3, 15, 27, 39, 51, \ldots\} = \bar{3}.$$

We have remarked in passing that abstraction often leads to greater insight. Our present approach to the concept of 'integers modulo n' has been direct but has taken no account of our existing knowledge. Suppose we reconsider 'integers modulo n' in the light of the group theory of the previous chapter. We see that $\mathbb{Z}$ is an additive group and that $n\mathbb{Z} = \{nk \,|\, k \in \mathbb{Z}\}$ is the cyclic subgroup generated by n; the equivalence classes modulo n are precisely the cosets of $n\mathbb{Z}$ in $\mathbb{Z}$. Now we may form the factor-group $\mathbb{Z}/n\mathbb{Z}$ consisting of these cosets as elements and with addition in $\mathbb{Z}/n\mathbb{Z}$ given by

$$\bar{a} + \bar{b} = \overline{a+b} \quad (a, b \in \mathbb{Z}).$$

Notation

The factor-group $\mathbb{Z}/n\mathbb{Z}$ is denoted by $\mathbb{Z}_n$, where for definiteness we suppose that $n > 1$.

Example 3

$$\mathbb{Z}_8 = \{\bar{0}, \bar{1}, \bar{2}, \bar{3}, \bar{4}, \bar{5}, \bar{6}, \bar{7}\}$$

and, for example,

$$\bar{2} + \bar{3} = \bar{5}, \quad \bar{4} + \bar{5} = \bar{9} = \bar{1},$$
$$\bar{5} + \bar{6} = \overline{11} = \bar{3}, \quad \bar{4} + \bar{4} = \bar{8} = \bar{0}.$$

What is perhaps less obvious from these considerations is that we may also introduce a multiplication with regard to the equivalence classes modulo n in such a way that $\mathbb{Z}_n$ becomes a ring. We summarise these facts in a theorem.

Theorem 1

Let n be a given integer, $n > 1$. Then $\mathbb{Z}_n$, the set of equivalence classes modulo n, is a commutative ring with an identity under the definitions:

$$\bar{a} + \bar{b} = \overline{a+b} \quad (a, b \in \mathbb{Z}),$$
$$\bar{a}\bar{b} = \overline{ab} \quad (a, b \in \mathbb{Z}).$$

Proof

We know that addition is unambiguously defined by $\bar{a} + \bar{b} = \overline{a+b}\,(a, b \in \mathbb{Z})$ and that, under this addition, $\mathbb{Z}_n$ is an additive group.

To establish the multiplication we must first show that the rule of multiplication above is meaningful. In other words if $\bar{a} = \bar{a'}$ and $\bar{b} = \bar{b'}$ $(a, b, a', b' \in \mathbb{Z})$ then we have to prove that $\overline{ab} = \overline{a'b'}$. But $\bar{a} = \bar{a'}$ implies that n divides $a - a'$ and, similarly, n divides $b - b'$. But

$$ab - a'b' = (a - a')b + a'(b - b')$$

and so n divides $ab - a'b'$, consequently $\overline{ab} = \overline{a'b'}$.

The associativity of the multiplication follows from the definition and the associativity of multiplication in $\mathbb{Z}$ since for $a, b, c \in \mathbb{Z}$ we have

$$(\bar{a}\bar{b})\bar{c} = (\overline{ab})\bar{c} = \overline{(ab)c} = \overline{a(bc)} = \bar{a}(\overline{bc}) = \bar{a}(\bar{b}\bar{c}).$$

In a similar manner the distributive laws hold since for $a, b, c \in \mathbb{Z}$ we have

$$\bar{a}(\bar{b} + \bar{c}) = \bar{a}(\overline{b+c}) = \overline{a(b+c)} = \overline{ab + ac} = \overline{ab} + \overline{ac} = \bar{a}\bar{b} + \bar{a}\bar{c}.$$

The other distributive law is equally easy to prove and so, finally, we conclude that $\mathbb{Z}_n$ is a ring. Fairly obviously $\mathbb{Z}_n$ is commutative and $\bar{1}$ is the identity since

$$\bar{1}\bar{a} = \overline{1a} = \bar{a} = \overline{a1} = \bar{a}\bar{1} \quad (a \in \mathbb{Z}). \qquad \square$$

Example 4

Following the above argument, let us write down the addition and multiplication tables of $\mathbb{Z}_6 = \{\bar{0}, \bar{1}, \bar{2}, \bar{3}, \bar{4}, \bar{5}\}$.

Sum	$\bar{0}$	$\bar{1}$	$\bar{2}$	$\bar{3}$	$\bar{4}$	$\bar{5}$
$\bar{0}$	$\bar{0}$	$\bar{1}$	$\bar{2}$	$\bar{3}$	$\bar{4}$	$\bar{5}$
$\bar{1}$	$\bar{1}$	$\bar{2}$	$\bar{3}$	$\bar{4}$	$\bar{5}$	$\bar{0}$
$\bar{2}$	$\bar{2}$	$\bar{3}$	$\bar{4}$	$\bar{5}$	$\bar{0}$	$\bar{1}$
$\bar{3}$	$\bar{3}$	$\bar{4}$	$\bar{5}$	$\bar{0}$	$\bar{1}$	$\bar{2}$
$\bar{4}$	$\bar{4}$	$\bar{5}$	$\bar{0}$	$\bar{1}$	$\bar{2}$	$\bar{3}$
$\bar{5}$	$\bar{5}$	$\bar{0}$	$\bar{1}$	$\bar{2}$	$\bar{3}$	$\bar{4}$

Product	$\bar{0}$	$\bar{1}$	$\bar{2}$	$\bar{3}$	$\bar{4}$	$\bar{5}$
$\bar{0}$	$\bar{0}$	$\bar{0}$	$\bar{0}$	$\bar{0}$	$\bar{0}$	$\bar{0}$
$\bar{1}$	$\bar{0}$	$\bar{1}$	$\bar{2}$	$\bar{3}$	$\bar{4}$	$\bar{5}$
$\bar{2}$	$\bar{0}$	$\bar{2}$	$\bar{4}$	$\bar{0}$	$\bar{2}$	$\bar{4}$
$\bar{3}$	$\bar{0}$	$\bar{3}$	$\bar{0}$	$\bar{3}$	$\bar{0}$	$\bar{3}$
$\bar{4}$	$\bar{0}$	$\bar{4}$	$\bar{2}$	$\bar{0}$	$\bar{4}$	$\bar{2}$
$\bar{5}$	$\bar{0}$	$\bar{5}$	$\bar{4}$	$\bar{3}$	$\bar{2}$	$\bar{1}$

Although $\mathbb{Z}_6$ is a commutative ring with $\bar{1}$ as identity, $\mathbb{Z}_6$ is not an integral domain since proper divisors of zero exist, for example, $\bar{2}.\bar{3} = \bar{0}$.

Example 5

Let us write down the addition and multiplication tables of $\mathbb{Z}_5 = \{\bar{0}, \bar{1}, \bar{2}, \bar{3}, \bar{4}\}$.

Sum	$\bar{0}$	$\bar{1}$	$\bar{2}$	$\bar{3}$	$\bar{4}$
$\bar{0}$	$\bar{0}$	$\bar{1}$	$\bar{2}$	$\bar{3}$	$\bar{4}$
$\bar{1}$	$\bar{1}$	$\bar{2}$	$\bar{3}$	$\bar{4}$	$\bar{0}$
$\bar{2}$	$\bar{2}$	$\bar{3}$	$\bar{4}$	$\bar{0}$	$\bar{1}$
$\bar{3}$	$\bar{3}$	$\bar{4}$	$\bar{0}$	$\bar{1}$	$\bar{2}$
$\bar{4}$	$\bar{4}$	$\bar{0}$	$\bar{1}$	$\bar{2}$	$\bar{3}$

Product	$\bar{0}$	$\bar{1}$	$\bar{2}$	$\bar{3}$	$\bar{4}$
$\bar{0}$	$\bar{0}$	$\bar{0}$	$\bar{0}$	$\bar{0}$	$\bar{0}$
$\bar{1}$	$\bar{0}$	$\bar{1}$	$\bar{2}$	$\bar{3}$	$\bar{4}$
$\bar{2}$	$\bar{0}$	$\bar{2}$	$\bar{4}$	$\bar{1}$	$\bar{3}$
$\bar{3}$	$\bar{0}$	$\bar{3}$	$\bar{1}$	$\bar{4}$	$\bar{2}$
$\bar{4}$	$\bar{0}$	$\bar{4}$	$\bar{3}$	$\bar{2}$	$\bar{1}$

In this case there are no proper divisors of zero and $\mathbb{Z}_5$ is an integral domain. We notice that, in fact, the set of non-zero elements $\{\bar{1}, \bar{2}, \bar{3}, \bar{4}\}$ is a cyclic group under multiplication.

Exercises 5.1

1. Write down the addition and multiplication tables of $\mathbb{Z}_2$, $\mathbb{Z}_3$ and $\mathbb{Z}_4$.

2. Find solutions to the following congruences:
$$4x \equiv 3 \ (\text{mod } 7), \quad 5y \equiv 4 \ (\text{mod } 11), \quad 2z \equiv 5 \ (\text{mod } 13).$$

3. Let m_1 and m_2 be coprime integers and let u_1 and u_2 be integers such that $u_1 m_1 + u_2 m_2 = 1$. For any $a_1, a_2 \in \mathbb{Z}$ prove that if $x = a_2 u_1 m_1 + a_1 u_2 m_2$ then $x \equiv a_1 \ (\text{mod } m_1)$ and $x \equiv a_2 \ (\text{mod } m_2)$. (Version of the Chinese Remainder Theorem, from China, date uncertain but before 636 AD.)

4. Let D and E be integral domains. Let R be the Cartesian product $D \times E$ and let an addition and multiplication be defined in R by

$$(d_1, e_1) + (d_2, e_2) = (d_1 + d_2, e_1 + e_2),$$

$$(d_1, e_1)(d_2, e_2) = (d_1 d_2, e_1 e_2) \quad (d_1, d_2 \in D, e_1, e_2 \in E).$$

Prove that R is a commutative ring with an identity but that R is not an integral domain.

5. Let D be an integral domain and let x be an element of D such that $x^2 = x$. Prove that either $x = 0$ or x is the identity of D.

6. Any modern book has an International Standard Book Number (ISBN) given in the form:

$$a_1 - a_2 a_3 a_4 - a_5 a_6 a_7 a_8 a_9 - a_{10}.$$

For example 'Finite-Dimensional Vector Spaces' by P.R. Halmos (Springer-Verlag, 1974) has ISBN 0-387-90093-4 and 'A History of Algebra' by B.L. van der Waerden (Springer-Verlag, 1985) has ISBN 0-387-13610-X (where X = 10). Verify that, in these two examples,

$$a_1 + 2a_2 + 3a_3 + 4a_4 + 5a_5 + 6a_6 + 7a_7 + 8a_8$$

$$+ 9a_9 + 10a_{10} \equiv 0 \ (\mathrm{mod}\ 11).$$

Try this congruence for the ISBN of any, randomly chosen, modern book.

5.2 Integral Domains and Fields

As we have just seen, the rings $\mathbb{Z}_5$ and $\mathbb{Z}_6$ have very different properties. We wish to characterize $\mathbb{Z}_n$, for general n, but first we require some definitions.

Definition 2

Let R be a ring with an identity 1. An element a of R is said to be **invertible**, or to be a **unit**, if there exists b in R such that $ab = ba = 1$. We write $b = a^{-1}$ and call b (if it exists) the **inverse** of a.

Examples 6

1. All non-zero elements of $\mathbb{Q}$, $\mathbb{R}$ and $\mathbb{C}$ are invertible.
2. $\mathbb{Z}$ has only two invertible elements, namely ± 1.

3. $\mathbb{Z}_6$ has two invertible elements $\bar{1}$ and $\bar{5}$.

4. $\mathbb{Z}_5$ has four invertible elements $\bar{1}, \bar{2}, \bar{3}$ and $\bar{4}$.

The existence, or otherwise, of invertible elements in a ring leads us to consider a 'field'. Dedekind gave in 1871 a definition of a field considered as a subset of the complex numbers, but the abstract definition of a field is due to Weber. A significant stimulus to research in fields was given in 1910 by E. Steinitz (1871–1928).

Definition 3

A commutative ring with an identity in which every non-zero element is invertible is called a **field** (German, *Körper*; French, *corps*).

Examples 7

1. $\mathbb{Q}$, $\mathbb{R}$ and $\mathbb{C}$ are fields.

2. $\mathbb{Z}_5$ is a field.

3. $\mathbb{Z}$ is not a field.

A convenient means of remembering the axioms of a field is given by the following characterization.

Theorem 2 Characterization of a Field

Let R be a non-empty set with two binary operations of addition and multiplication such that the following hold.

1. R is an additive Abelian group.

2. $R \setminus \{0\}$ is a multiplicative Abelian group.

3. The distributive laws hold.

Then R is a field.

Theorem 3

Every field is an integral domain.

Proof

Let F be a field. Then F is a commutative ring with an identity 1. We have only to show that F has no proper divisors of zero. Suppose there exist $a, b \in F$ such that $ab = 0$. Now as $a \neq 0$ there exists a^{-1} such that $a^{-1}a = 1$ and so we may write $b = 1b = (a^{-1}a)b = a^{-1}(ab) = a^{-1}0 = 0$. Similarly if $b \neq 0$ we prove that $a = 0$. Hence we deduce that F is an integral domain.   □

As we observed above, an infinite integral domain, such as $\mathbf{Z}$, need not be a field. However, for finite integral domains, we have an explicit result.

Theorem 4

A finite integral domain is a field.

Proof

Let D be a finite integral domain. To prove that D is a field we have to show that $D \setminus \{0\}$ is multiplicatively a group. Certainly $D \setminus \{0\}$ is closed under multiplication since $x, y \in D \setminus \{0\}$ implies that x and y are not proper divisors of zero and so $xy \neq 0$ and hence $xy \in D \setminus \{0\}$. Associativity being evident, $D \setminus \{0\}$ is a semigroup. We claim that $D \setminus \{0\}$ admits cancellation. Thus suppose there exist $a, b, x \in D \setminus \{0\}$ and $ax = bx$. By Example 10 in Chapter 3, we deduce that $a = b$ and so we have cancellation. We have already shown that a finite semigroup with cancellation is a group (Theorem 3, Chapter 4) and so $D \setminus \{0\}$ is a group and then D is a field.   □

Theorem 5

$\mathbf{Z}_p$ is a field if and only if p is a prime.

Proof

Suppose p is not a prime. We show that proper divisors of zero exist in $\mathbf{Z}_n$. We have $p = n_1 n_2$ for some $n_1, n_2, \in \mathbf{N}$ where $1 < n_1 < n$, $1 < n_2 < n$. Then $\overline{n_1} \neq \overline{0}$, $\overline{n_2} \neq \overline{0}$ but $\overline{n_1}\,\overline{n_2} = \overline{n_1 n_2} = \overline{p} = \overline{0}$. Thus $\mathbf{Z}_p$ is not an integral domain.

Suppose now that n is a prime. We may prove that $\mathbf{Z}_p$ is a field by reversing the argument of the previous paragraph to show that $\mathbf{Z}_p$ is an integral domain and then, from Theorem 4, we deduce that $\mathbf{Z}_p$ is a field. However, it is perhaps more instructive to prove the result directly without invoking Theorem 4. We know that $\mathbf{Z}_p$ is a commutative ring with an identity $\overline{1}$. We have to show therefore that the non-zero elements of $\mathbf{Z}_p$ are invertible. Now we have $\mathbf{Z}_p = \{\bar{0}, \bar{1}, \ldots, \overline{p-1}\}$. Let $\bar{a}$ be a non-zero element of $\mathbf{Z}_p$, $a \in \{1, 2, \ldots, p-1\}$. Then as p is a prime, a and p are coprime and, by the Euclidean Algorithm, there exist $x, y \in \mathbf{Z}$ such that $ax + py = 1$. Hence

$$\bar{1} = \overline{ax + py} = \overline{ax} + \overline{py} = \bar{a}\bar{x} + \bar{p}\bar{y} = \bar{a}\bar{x} + \bar{0}\bar{y} = \bar{a}\bar{x}$$

and so $\bar{a}$ has $\bar{x}$ as inverse. Hence finally $\mathbf{Z}_p$ is a field. $\qquad\square$

Since $\mathbf{Z}_p$, p prime, is a field, $\mathbf{Z}_p$ has some features in common with $\mathbf{Q}$ and $\mathbf{R}$. Some algebraic manipulations which are familiar in $\mathbf{Q}$ and $\mathbf{R}$ may also be performed in $\mathbf{Z}_p$.

Examples 8

Suppose we try to solve, if possible, the following pairs of simultaneous linear equations over $\mathbf{Z}_7$.

1.
$$\left.\begin{array}{l} \bar{2}x + \bar{3}y = \bar{1} \\ \bar{3}x + y = \bar{4} \end{array}\right\}.$$

We make the coefficients of x the same in both equations by multiplying the first by $\bar{3}$ and the second by $\bar{2}$. Thus we obtain

$$\left.\begin{array}{l} \bar{3}.\bar{2}x + \bar{3}.\bar{3}y = \bar{3}.\bar{1} \\ \bar{2}.\bar{3}x + \bar{2}y = \bar{2}.\bar{4} \end{array}\right\}$$

giving

$$\left.\begin{array}{l} \bar{6}x + \bar{2}y = \bar{3} \\ \bar{6}x + \bar{2}y = \bar{1} \end{array}\right\}.$$

These equations are therefore inconsistent and no solution is possible.

2.
$$\left.\begin{array}{l} \bar{3}x + \bar{6}y = \bar{6} \\ \bar{4}x + \bar{5}y = \bar{4} \end{array}\right\}.$$

We then have

$$\left.\begin{array}{l} \bar{4}.\bar{3}x + \bar{4}.\bar{6}y = \bar{4}.\bar{6} \\ \bar{3}.\bar{4}x + \bar{3}.\bar{5}y = \bar{3}.\bar{4} \end{array}\right\}$$

giving

$$\left.\begin{array}{l} \bar{5}x + \bar{3}y = \bar{3} \\ \bar{5}x + y = \bar{5} \end{array}\right\}.$$

Subtracting we have

$$\bar{2}y = \bar{3} - \bar{5} = -\bar{2} = \bar{5}.$$

So far we have used only ring-theoretic properties of $\mathbb{Z}_7$. Now we use the invertibility of non-zero elements of $\mathbb{Z}_7$. $\bar{2}$ has $\bar{4}$ as an inverse and so we have

$$\bar{4}.\bar{2}y = \bar{4}.\bar{5}$$

which yields

$$y = \bar{1}y = \bar{6}.$$

Similarly we may show that $x = \bar{4}$.

In the ring of integers $\mathbb{Z}$ we may form the sums

$$1, 1+1, 1+1+1, 1+1+1+1, \ \ldots$$

and these sums are distinct, being respectively $1, 2, 3, 4, \ldots$ On the other hand, in $\mathbb{Z}_n$ the corresponding sequence

$$\bar{1}, \bar{1}+\bar{1}, \bar{1}+\bar{1}+\bar{1}, \bar{1}+\bar{1}+\bar{1}+\bar{1}, \ \ldots$$

repeats as we obtain

$$\bar{1}, \bar{2}, \ldots, \overline{n-1}, \bar{0}, \bar{1}, \bar{2}, \ \ldots, \overline{n-1}, \bar{0}, \ \ldots$$

In the first case there is no integer n ($n > 0$) for which $n1 = 0$ but in the second case $n\bar{1} = \bar{0}$. These two cases require to be appropriately distinguished.

Definition 4

Let R be a ring with an identity 1. R is said to have **finite characteristic** if there exists an integer n ($n > 0$) such that

$$0 = 1+1+\ldots+1 \quad (n \text{ terms})$$

or, equivalently, that $0 = n1$. Otherwise R is said to have **infinite** or **zero** characteristic. (It may seem to be a trifle odd that in this situation the words 'infinite' and 'zero' are regarded as synonymous but such indeed is the case.)

Before relating this definition to integral domains we prove an interesting ancillary result.

Lemma 2

Let R be a ring with an identity 1. Let R have finite characteristic and suppose that $n1 = 0$ ($n > 0$). Then for all elements a of R, $na = 0$.

Proof

$$na = a + a + \ldots + a \quad (n \text{ terms})$$
$$= 1a + 1a + \ldots + 1a$$
$$= (1 + 1 + \ldots + 1)a \quad (\text{distributivity})$$
$$= (n1)a = 0a = 0. \qquad \square$$

Our main interest is in the characteristics of integral domains and fields.

Theorem 6

Let D be an integral domain with identity 1 and of finite characteristic. Then there exists a unique prime p such that $p1 = 0$.

Proof

By assumption there exists an integer n $(n > 0)$ such that $n1 = 0$. Let p be chosen to be the least such integer, $p1 = 0$. We claim that p is a prime. Suppose, for the sake of argument, that p is not a prime and let $p = p_1 p_2$ where

$$1 < p_1 < p, 1 < p_2 < p \quad (p_1, p_2 \in \mathbb{N}).$$

Then

$$(p_1 1)(p_2 1) = (1 + 1 + \ldots + 1)(1 + 1 + \ldots + 1)$$

where we have p_1 terms in the first bracket and p_2 terms in the second. Expanding by distributivity and collecting terms we have $p_1 p_2$ terms of the form $11 = 1$. Thus

$$(p_1 1)(p_2 1) = (p_1 p_2)1 = p1 = 0.$$

But D is an integral domain and so $p_1 1 = 0$ or $p_2 1 = 0$. But either conclusion contradicts the choice of p as least integer such that $p1 = 0$. Hence p is a prime and is unique. $\qquad \square$

Definition 5

Let D be an integral domain with identity 1 and of finite characteristic. The least integer p such that $p1 = 0$ is a prime called the **characteristic** of D. D is said to be of **prime characteristic**.

Examples 9

1. $\mathbb{Z}$, $\mathbb{Q}$, $\mathbb{R}$, $\mathbb{C}$ are of infinite characteristic. For each prime p, $\mathbb{Z}_p$ is of prime characteristic p.

2. In a field F of characteristic 2, addition and subtraction are the same! This apparently paradoxical result arises as follows. We have $2a = 0$ and so $a + a = 0$. But then $a = -a$. Hence $b + a = b - a$ for all $a, b \in F$.

3. A field which is finite necessarily has finite characteristic. But a finite field of prime characteristic p need not have p elements. The following tables exhibit a field E of characteristic 2 having four elements 0, 1, α, β.

Sum	0	1	α	β
0	0	1	α	β
1	1	0	β	α
α	α	β	0	1
β	β	α	1	0

Product	0	1	α	β
0	0	0	0	0
1	0	1	α	β
α	0	α	β	1
β	0	β	1	α

We notice that $F = \{0, 1\}$ is a 'subfield' of E. The quadratic polynomial $x^2 + x + 1$ does not factorize in $F[x]$ but does factorize in $E[x]$ since

$$x^2 + x + 1 = x^2 + (\alpha + \beta)x + \alpha\beta = (x + \alpha)(x + \beta).$$

We may also notice that $E \setminus \{0\} = \{1, \alpha, \beta\}$ is multiplicatively a cyclic group of order 3 generated by either α or β.

Definition 6

Let F be a non-empty subset of the field E which is also a field under the addition and multiplication in E. Then F is called a **subfield** of E.

Theorem 7 Subfield Criterion

A non-empty subset F of a field E is a subfield of E if and only if the following axioms of a field are satisfied.

1. (i) F is closed under addition.
   (iv) For all $a \in F$, $-a \in F$.

2. (i) F is closed under multiplication.
   (iii) $1 \in F$.
   (iv) For all $a \in F$, $a \neq 0$, $a^{-1} \in F$.

Proof

Certainly if F is a subfield the axioms above must be satisfied. On the other hand if the axioms above are satisfied then F is a subring and the remaining axioms for a field are satisfied in F since they are satisfied in E. Hence we conclude that F is a subfield of E. □

In certain instances Theorem 7 may obviate some lengthy verifications of the axioms for a field. If we know that an algebraic structure is a subset of a field we may be able to establish more easily that the structure is actually a subfield.

Example 10

Let $F = \{a + b\sqrt{2} \mid a, b \in \mathbb{Q}\}$. Then F is a non-empty subset of the field $\mathbb{R}$. To prove that F is a subfield we may utilize Theorem 7.

1. (i)   Typical elements of F are $a + b\sqrt{2}$ and $c + d\sqrt{2}$ $(a, b, c, d \in \mathbb{Q})$ and their sum is

$$(a + b\sqrt{2}) + (c + d\sqrt{2}) = (a + c) + (b + d)\sqrt{2}$$

which is of the form $q_1 + q_2\sqrt{2}$ where $q_1, q_2 \in \mathbb{Q}$. Thus addition in F is closed.

   (iv) $a + b\sqrt{2} \in F$ implies $-(a + b\sqrt{2}) = (-a) + (-b)\sqrt{2} \in F$.

2. (i)   Let $a, b, c, d \in \mathbb{Q}$. Then

$$(a + b\sqrt{2})(c + d\sqrt{2}) = (ac + 2bd) + (ad + bc)\sqrt{2} \in F$$

since $ac + 2bd, ad + bc \in \mathbb{Q}$. Thus multiplication in F is closed.

   (iii) $1 = 1 + 0\sqrt{2} \in F$.

   (iv) Let $a + b\sqrt{2} \neq 0$ $(a, b \in \mathbb{Q})$. Then a and b are not both zero. But then $a - b\sqrt{2} \neq 0$ since $a - b\sqrt{2} = 0$ would imply that $a = b = 0$ since we have earlier shown that $\sqrt{2} \notin \mathbb{Q}$. Then

$$(a + b\sqrt{2})^{-1} = \frac{1}{a + b\sqrt{2}} = \frac{a - b\sqrt{2}}{(a + b\sqrt{2})(a - b\sqrt{2})}$$

$$= \frac{a - b\sqrt{2}}{a^2 - 2b^2} = \frac{a}{a^2 - 2b^2} + \frac{(-b)}{a^2 - 2b^2}\sqrt{2} \in F.$$

Thus, finally, we conclude that F is a subfield of $\mathbb{R}$.

Integral domains of finite characteristic have a version of the well-known binomial theorem which, on first encounter, is mildly amusing. We illustrate this first by means of examples.

Examples 11

1. Let D be an integral domain of characteristic 2. Let $a, b \in D$. Then, as D is commutative, we may expand $(a + b)^2$ to give

$$(a + b)^2 = a^2 + ab + ba + b^2 = a^2 + 2ab + b^2 = a^2 + b^2 \text{ since } 2ab = 0.$$

2. Let D be an integral domain of characteristic 3. Let $a, b \in D$. Then, as D is commutative, we may expand $(a + b)^3$ to give

$$(a + b)^3 = a^3 + 3a^2b + 3ab^2 + b^3 = a^3 + b^3.$$

These examples should suggest a general result which we now prove.

Theorem 8

Let D be an integral domain of prime characteristic p. Let $a, b \in D$. Then

$$(a + b)^p = a^p + b^p.$$

Proof

We have to use the so-called binomial theorem, namely

$$(a + b)^p = a^p + pa^{p-1}b + \frac{p(p - 1)}{1.2} a^{p-2}b^2 + \ldots + b^p.$$

A typical term in this expansion, other than the first or the last, is

$$\frac{p(p - 1) \ldots (p - r + 1)}{1.2 \ldots r} a^{p-r}b^r$$

where $1 \leq r \leq p - 1$. Now

$$\frac{p(p - 1) \ldots (p - r + 1)}{1.2 \ldots r}$$

is a strictly positive integer and so $1.2 \ldots r$ must divide $p(p - 1) \ldots (p - r + 1)$. But p is a prime and $p > r$ so none of $1, 2, \ldots, r$ can divide p but each must divide the product $(p - 1)(p - 2) \ldots (p - r + 1)$. In consequence $\frac{(p - 1) \ldots (p - r + 1)}{1.2 \ldots r}$ is an integer. Thus p divides

$$\frac{p(p - 1) \ldots (p - r + 1)}{1.2 \ldots r}$$

Hence

$$\frac{p(p - 1) \ldots (p - r + 1)}{1.2 \ldots r} a^{p-r}b^r = 0.$$

Thus, finally,

$$(a + b)^p = a^p + b^p.$$ □

Corollary

Let D be an integral domain of prime characteristic p. Let $a_1, a_2, \ldots, a_n$ be elements of $\dot{D}$. Then

$$(a_1 + a_2 + \ldots + a_n)^p = a_1^p + a_2^p + \ldots + a_n^p.$$

Proof

A simple induction argument gives the result. Thus suppose

$$(a_1 + a_2 + \ldots + a_k)^p = a_1^p + a_2^p + \ldots + a_k^p.$$

for some $k \geq 2$. Then

$$(a_1 + a_2 + \ldots + a_{k+1})^p = [(a_1 + \ldots + a_k) + a_{k+1}]^p = (a_1 + \ldots + a_k)^p + a_{k+1}^p$$

$$= a_1^p + a_2^p + \ldots + a_k^p + a_{k+1}^p.$$

The result is immediate. □

We conclude this section with a result drawing together integral domains and fields. A finite integral domain is a field but an infinite integral domain may or may not be a field. Now from the integral domain of the integers $\mathbb{Z}$ we obtain the quotient field $\mathbb{Q}$ of $\mathbb{Z}$ by simply taking quotients of elements of $\mathbb{Z}$ to form the elements of $\mathbb{Q}$; in a similar way we may form the 'quotient field' of any integral domain.

From the numbers 1, 2, 3, 4, ... we form quotients such as $\frac{1}{2}, \frac{3}{4}, \frac{2}{4}, \ldots$ where we identify $\frac{1}{2}, \frac{2}{4}, \frac{3}{6}, \ldots$ and 2, $\frac{2}{1}, \frac{4}{2}, \ldots$ etc. As we are long accustomed to make such identifications no difficulty is experienced in their manipulation. On reflection we realise that we are regarding $\frac{1}{2}, \frac{2}{4}, \frac{3}{6}, \ldots$ as equivalent under some unstated equivalence relation. If, therefore, we are to consider an arbitrary integral domain, we must make more precise the nature of an implicit equivalence relation. Instead of 'quotients' of two elements we begin more properly with ordered pairs of elements and then we introduce an appropriate equivalence relation. From a given integral domain D we construct a field F in which the integral domain is 'embedded', that is there is a bijective mapping of D into F preserving addition and multiplication, for example $\mathbb{Z}$ is 'embedded' in $\mathbb{Q}$. (A more precise definition of embedding will be given in a subsequent section.)

Theorem 9

Let D be an integral domain. Then D may be embedded in a field.

Proof

Let $G = D \times (D \setminus \{0\})$. Then G is the set of ordered pairs (a, b) where a, b are elements of D, $b \neq 0$. (We shall construct a field consisting of what, informally, are quotients $\frac{a}{b}$. The proof will perhaps be more easily comprehended if the reader thinks of (a, b) as standing for $\frac{a}{b}$.)

The proof is in three steps.

Step 1 We claim that the relation $\sim$ on G given by $(a, b) \sim (c, d)$ if and only if $ad = bc$ is an equivalence relation. Certainly $(a, b) \sim (a, b)$ and if $(a, b) \sim (c, d)$ we have $ad = bc$, giving $cb = da$ and so $(c, d) \sim (a, b)$. Suppose we have $(a, b) \sim (c, d)$ and $(c, d) \sim (e, f)$ then $ad = bc$ and $cf = ed$. Then using commutativity and associativity we have

$$(af)d = a(fd) = a(df) = (ad)f = (bc)f = b(cf) = b(ed) = (be)d.$$

Since D is an integral domain and $d \neq 0$ we have $af = be$ and so $(a, b) \sim (e, f)$.

Step 2 We let F be the set of equivalence classes of G under $\sim$. We let $\overline{(a, b)}$ denote the equivalence class to which (a, b) belongs. We introduce an addition and multiplication on F and claim that these operations are well-defined. We have to show that if $\overline{(a, b)} = \overline{(a', b')}$ and $\overline{(c, d)} = \overline{(c', d')}$ then

$$\overline{(ad + bc, bd)} = \overline{(a'd' + b'c', b'd')}$$

and

$$\overline{(ac, bd)} = \overline{(a'c', b'd')}.$$

But

$$ab' = ba' \text{ and } cd' = dc'$$

and so

$$(a'd' + b'c')(bd) = (a'd')(bd) + (b'c')(bd)$$
$$= (a'b)(dd') + (c'd)(bb')$$
$$= (ab')(dd') + (cd')(bb')$$
$$= (ad)(b'd') + (bc)(b'd')$$
$$= (ad + bc)(b'd')$$

and

$$(ac)(b'd') = (ab')(cd') = (ba')(dc') = (a'c')(bd).$$

Intuitively

$$\frac{a}{b} = \frac{a'}{b'}, \frac{c}{d} = \frac{c'}{d'} \text{ implies } \frac{a}{b} + \frac{c}{d} = \frac{a'}{b'} + \frac{c'}{d'}$$

from which

$$\frac{ad+bc}{bd} = \frac{a'd'+b'c'}{b'd'}. \text{ Also } \frac{a}{b} \cdot \frac{c}{d} = \frac{a'}{b'} \cdot \frac{c'}{d'}.$$

Hence $(ad+bc, bd) \sim (a'd'+b'c', b'd')$ and $(ac, bd) \sim (a'c', b'd')$ giving the result. Thus our definitions are well-founded.

Step 3 We prove that F is a field. Addition is closed and also associative since

$$[(\overline{a,b})(\overline{c,d})] + (\overline{e,f}) = (\overline{ad+bc, bd}) + (\overline{e,f})$$

$$= (\overline{adf+bcf+bde, bdf})$$

$$= (\overline{a,b}) + (\overline{cf+de, df})$$

$$= (\overline{a,b}) + [(\overline{c,d})(\overline{e,f})].$$

The identity element is $(\overline{1,1}) = (\overline{x,x})$ for all $x \in D$, $x \neq 0$. We have to show that if $(\overline{a,b}) \neq (\overline{0,1})$ then $(\overline{a,b})$ has an inverse. Since $(\overline{a,b}) \neq (\overline{0,1})$ we have $a = a1 \neq b0 = 0$ and so $(b,a) \in D \times (D \setminus \{0\})$. Then $(\overline{a,b})(\overline{b,a}) = (\overline{ab, ab}) = (\overline{1,1})$. Similarly $(\overline{b,a})(\overline{a,b}) = (\overline{1,1})$. Multiplication is commutative. It may be shown that the distributive laws hold and so F is a field.

We may also show that the subset of F consisting of the elements $(\overline{x,1})$ $(x \in \mathbb{D})$ is a sub-integral domain of D and that the mapping

$$x \rightarrow (\overline{x,1}) \quad (x \in D) \quad \left[\text{Intuitively } x \rightarrow \frac{x}{1}.\right]$$

embeds D in F. □

Definition 7

The field F, constructed in Theorem 9, is called the **quotient field** of D.

Remark

From $\mathbb{Z}$ we construct $\mathbb{Q}$ as the quotient field of $\mathbb{Z}$. From the polynomial domain $\mathbb{Q}[x]$ we construct $\mathbb{Q}(x)$ (note round brackets) as the field of 'rational functions in x' over $\mathbb{Q}$.

Exercises 5.2

1. In $\mathbb{Z}_7$ find the roots of the following polynomials and factorize the polynomials as fully as possible:

$$x^2 + x + \overline{2}, \quad x^2 + \overline{2}x + \overline{4}, \quad x^2 + x + \overline{3}, \quad x^6 - \overline{1}.$$

2. In $\mathbb{Z}_7$ find the inverses of $\overline{2}, \overline{3}$ and $\overline{6}$.

3. In $\mathbb{Z}_{47}$ find the inverses of $\overline{3}, \overline{23}, \overline{24}$ and $\overline{32}$.

4. Over $\mathbb{Z}_5$ solve, if possible, the following pairs of simultaneous equations.

   (i) $\begin{aligned} \overline{2}x + \overline{3}y &= \overline{1}, \\ \overline{3}x + \overline{2}y &= \overline{4}. \end{aligned}$

   (ii) $\begin{aligned} x + \overline{3}y &= \overline{2}, \\ \overline{3}x + \overline{2}y &= \overline{2}. \end{aligned}$

5. Over $\mathbb{Z}_{31}$ solve, if possible, the following simultaneous equations.

$$\overline{15}x + \overline{25}y = \overline{16},$$
$$\overline{8}x + \overline{21}y = \overline{18}.$$

6. For the field F, given as Example 9 no. 3 above and consisting of the four elements $0, 1, \alpha, \beta$, solve the following simultaneous equations.

$$\alpha x + y = 1,$$
$$x + \beta y = \beta.$$

7. Prove that if R is a ring with an identity then the subset $U(R)$ of invertible elements of R is multiplicatively a group.

8. Let F be a field and x an indeterminate. What are the invertible elements of $F[x]$?

9. Let S be the subset $\{a + b\sqrt{3} \,|\, a, b \in \mathbb{Z}\}$ of $\mathbb{R}$. Prove that S is an integral domain. Is S a field?

10. Let S be the subset $\{a + b\sqrt{5}\,|\,a, b \in \mathbb{Q}\}$ of $\mathbb{R}$. Prove that S is a field.

11. Let S be the subset $\{a + b\sqrt{6}\,|\,a, b \in \mathbb{Q}\}$. Is S a field?

12. Let D be an integral domain of prime characteristic p. Prove that, for all $a, b \in D$ and $n \in \mathbb{N}$,

$$(a + b)^{p^n} = a^{p^n} + b^{p^n}.$$

Generalize.

13. Let Q be the set of 2×2 matrices of the form

$$\begin{pmatrix} a + ib & c + id \\ -c + id & a - ib \end{pmatrix} \quad (a, b, c, d \in \mathbb{R}).$$

Prove that Q is a ring which satisfies all the requirements of a field except commutativity of multiplication. (Q is the ring of the **quaternions**, first discovered by W. R. Hamilton, 1805–65.)

5.3 Euclidean Domains

We have shown that the integral domain of the integers $\mathbb{Z}$ and the polynomial domain $\mathbb{Q}[x]$ both have division and Euclidean algorithms. Here we study certain integral domains which are assumed to admit a division algorithm from which a Euclidean algorithm is naturally derivable.

Definition 8

Let D be an integral domain. Let there exist a mapping

$$v : D \setminus \{0\} \to \{0, 1, 2, \ldots\}$$

with the following two properties.

1. For all $a, b \in D \setminus \{0\}, v(a) \le (ab)$.
2. For all $a, b \in D \setminus \{0\}$, there exist $q, r \in D$ such that $a = bq + r$ where either $r = 0$ or, if $r \ne 0$, then $v(r) < v(b)$.

Then D with the mapping v is called a **Euclidean domain**.

Examples 12

1. $\mathbb{Z}$ with the mapping v defined by $v(a) = |a|$ $(a \in \mathbb{Z}, a \ne 0)$ is a Euclidean domain. To see this, we observe

(i)   that

$$v(ab) = |ab| = |a||b| \geq |a| = v(a) \quad (a, b \in \mathbb{Z}, a, b \neq 0)$$

since $v(a) \geq 1$, and that

(ii)   becomes simply a restatement of the division algorithm for $\mathbb{Z}$.

2. $F[x]$, where F is any field, with the mapping v defined by

$$v(f(x)) = \deg f(x) \quad (f(x) \in F[x], f(x) \neq 0)$$

is a Euclidean domain. To see this, we observe first that for

$$f(x), g(x) \in F[x], f(x) \neq 0, g(x) \neq 0,$$

$$v(f(x)g(x)) = \deg (f(x)g(x)) = \deg f(x) + \deg g(x) \geq \deg f(x) = v(f(x)).$$

Again 2, in Definition 9, is the division algorithm for $F[x]$. Strictly we have only discussed the division algorithm for polynomial domains over $\mathbb{Q}$ or $\mathbb{R}$ but the arguments previously employed for these particular fields easily adapt to any field. We illustrate some of these arguments in the next example.

Example 13

Consider $\mathbb{Z}_5[x]$. Let $a(x) = \bar{3}x^4 + \bar{2}x^3 + \bar{2}x^2 + x + \bar{3}$ and $b(a) = x^2 + \bar{2}$. Suppose we want to find $q(x), r(x) \in \mathbb{Z}_5[x]$ as in the definition of a Euclidean domain. We perform a long division as follows:

$$
\begin{array}{r}
3x^2 + \bar{2}x + \bar{1} \\
x^2 + \bar{2} \overline{) \bar{3}x^4 + \bar{2}x^3 + \bar{2}x^2 + x + \bar{3}} \\
\underline{\bar{3}x^4 \qquad\quad + x^2} \\
\bar{2}x^3 + x^2 + x \\
\underline{\bar{2}x^3 \qquad + \bar{4}x} \\
x^2 + \bar{2}x + \bar{3} \\
\underline{x^2 \qquad\quad + \bar{2}} \\
\bar{2}x + \bar{1}
\end{array}
$$

Thus $q(x) = \bar{3}x^2 + \bar{2}x + \bar{1}$, $r(x) = \bar{2}x + \bar{1}$.

Before considering other Euclidean domains we prove a convenient lemma. Recall that an element of a ring is a unit if and only if it is invertible.

Lemma 3

Let D be a Euclidean domain with identity 1, as in Definition 8.

1. For all $a \in D \setminus \{0\}, v(1) \leq v(a)$.
2. For $a \in D \setminus \{0\}, v(1) = v(a)$ if and only if a is a unit of D.

Proof

1. $v(1) \leq v(1a) = v(a) \ (a \in D, a \neq 0)$.
2. Suppose a is a unit. Then there exists $b \in D$ such that $ab = 1$. Then we have
   $v(a) \leq v(ab) = v(1) \leq v(a)$ and so $v(1) = v(a)$.

Conversely suppose $v(a) = v(1)$. There exist $q, r \in D$ such that $1 = aq + r$, where
if $r \neq 0$ we have $v(r) < v(a) = v(1)$ which is false and so $r = 0$. But, then, $aq = 1$
and a is a unit. $\square$

Our next example of a Euclidean domain occurs as a subintegral domain of the
field of complex numbers.

We recall that for complex numbers z_1 and z_2 we have

$$|z_1 z_2| = |z_1| |z_2|,$$
$$|z_1 + z_2| \leq |z_1| + |z_2|.$$

Lemma 4

Let $\mathbb{Z}[i]$ denote the set of complex numbers of the form $m + ni \ (m, n \in \mathbb{Z})$. Then
$\mathbb{Z}[i]$ is an integral domain.

Proof

$\mathbb{Z}[i]$ is a non-empty subset of the field $\mathbb{C}$. Now for $m_1, m_2, n_1, n_2 \in \mathbb{Z}$

$$(m_1 + n_1 i) + (m_2 + n_2 i) = (m_1 + m_2) + (n_1 + n_2)i \in \mathbb{Z}[i],$$
$$-(m_1 + n_1 i) = (-m_1) + (-n_1)i \in \mathbb{Z}[i],$$
$$(m_1 + n_1 i)(m_2 + n_2 i) = (m_1 m_2 - n_1 n_2) + (m_1 n_2 + n_1 m_2)i \in \mathbb{Z}[i].$$

The remaining axioms for an integral domain are evidently satisfied. $\square$

Definition 9

A complex number of the form $m + ni$ $(m, n \in \mathbb{Z})$ is called a **Gaussian integer** (after K.F. Gauss).

Theorem 10

The Gaussian integers $\mathbb{Z}[i]$ form a Euclidean domain under the mapping v given by

$$v(a) = |a|^2 \quad (a \in \mathbb{Z}[i], a \neq 0).$$

Proof

1. We want to prove that

$$v(a) \leq v(ab) \quad (a, b, \in \mathbb{Z}[i] \setminus \{0\}).$$

We note first that $b = b_1 + b_2 i$ where $b_1, b_2 \in \mathbb{Z}$ and so $v(b) = |b| = b_1^2 + b_2^2 \geq 1$. Hence $v(a) = |a|^2 \leq |a|^2 |b|^2 = |ab|^2 = v(ab)$.

2. Given $a, b \in \mathbb{Z}[i] \setminus \{0\}$, we have to find suitable q, r such that $a = bq + r$. Now $\dfrac{a}{b}$ is a complex number which must be of the form $\alpha + \beta i$ where $\alpha, \beta \in \mathbb{Q}$.

We now use a fairly obvious property of $\mathbb{Q}$. Since α must lie between two consecutive integers, we may therefore choose that integer m for which $\alpha = m + \varepsilon$, $|\varepsilon| < \frac{1}{2}$ (for example, $\alpha = \frac{18}{5}$ lies between 3 and 4 and we choose $m = 4, \varepsilon = -\frac{2}{5}$.) Similarly we choose n so that $\beta = n + \eta$, $|\eta| < \frac{1}{2}$. Let $q = m + ni$. Then $q \in \mathbb{Z}[i]$. Let $r = a - bq$. Then $r \in \mathbb{Z}[i]$. Thus

$$\frac{a}{b} = \alpha + \beta i = (m + \varepsilon) + (n + \eta)i = (m + ni) + (\varepsilon + \eta i) = q + (\varepsilon + \eta i)$$

and hence

$$r = a - bq = bq + b(\varepsilon + \eta i) - bq = b(\varepsilon + \eta i).$$

If $r = 0$ the appropriate q, r have been found. If $r \neq 0$ then

$$v(r) = |b(\varepsilon + \eta i)|^2 = |b|^2 |\varepsilon + \eta i|^2 = |b|^2 (\varepsilon^2 + \eta^2) \leq |b|^2 \left(\frac{1}{4} + \frac{1}{4} \right)$$

$$= \frac{1}{2} |b|^2 < |b|^2 = v(b).$$

Hence we have established that $\mathbb{Z}[i]$ is a Euclidean domain.　　□

Example 14

Let $a = 5 + 6i$, $b = 2 - 3i$. Suppose we have to find suitable $q, r \in \mathbb{Z}[i]$ such that $a = bq + r$ where either $r = 0$ or if $r \neq 0$ then $|r| < |b| = \sqrt{2^2 + 3^2} = \sqrt{13}$. Now

$$\frac{a}{b} = \frac{5 + 6i}{2 - 3i} = \frac{(5 + 6i)(2 + 3i)}{(2 - 3i)(2 + 3i)} = \frac{-8 + 27i}{13} = -\frac{8}{13} + \frac{27}{13}i.$$

Choosing the integers nearest to $-\frac{8}{13}$ and $\frac{27}{13}$ we obtain, respectively, -1 and 2. Let $q = -1 + 2i$ and $r = a - bq = (5 + 6i) - (2 - 3i)(-1 + 2i) = (5 + 6i) - (4 + 7i) = 1 - i$. We note that $|r| = \sqrt{1^2 + 1^2} = \sqrt{2} < \sqrt{13}$. Thus we have found an appropriate q, r. However, we may also note that q, r are not uniquely determined, for if $q' = 2i$ and $r' = a - bq' = (5 + 6i) - (2 - 3i)(2i) = -1 + 2i$ then $|r'| = \sqrt{5}$ and so q' and r' also satisfy the required conditions.

Exercises 5.3

1. Find the units of the Euclidean domain $\mathbb{Z}[i]$.

2. In the Euclidean domain $\mathbb{Z}[i]$ for given $a, b \in \mathbb{Z}[i]$ find $q, r \in \mathbb{Z}[i]$ such that $a = bq + r$ where $0 \leq |r| < |b|$:

   (i) $a = 1 + 13i, b = 4 - 3i$,

   (ii) $a = 5 + 15i, b = 7 + i$,

   (iii) $a = 5 + 6i, b = 2 - 3i$.

3. Let $D = \{\alpha + \beta\sqrt{2} \,|\, \alpha, \beta \in \mathbb{Z}\}$. Prove that D is a Euclidean domain if $v : D \setminus \{0\} \rightarrow \{0, 1, 2, \ldots\}$ is given by

$$v(\alpha + \beta\sqrt{2}) = |(\alpha - \beta\sqrt{2})(\alpha + \beta\sqrt{2})| = |\alpha^2 - 2\beta^2| \ (\alpha, \beta \in \mathbb{Z}).$$

4. (Hard) Let w be a non-trivial complex cube root of 1. Then $w^3 = 1$, $w \neq 1$ and so $w^2 + w + 1 = 0$ and the complex conjugate of w is $\overline{w} = w^2 = \dfrac{1}{w}$. (For the considerations of this example it is best to ignore the fact that $w = \dfrac{-1 \pm i\sqrt{3}}{2}$.)

   Prove that $D = \{\alpha + \beta w \,|\, \alpha, \beta \in \mathbb{Z}\}$ is a Euclidean domain if $v : D \setminus \{0\} \rightarrow \{0, 1, 2, \ldots\}$ is given by

$$v(\alpha + \beta w) = |\alpha + \beta w|^2 = (\alpha + \beta w)(\alpha + \beta\overline{w}).$$

5.4 Ideals and Homomorphisms

P. Fermat (1601–65), in commenting on the 'Arithmetica' of Diophantus (c. 250 AD), formulated a statement, of which he claimed to have a proof, that the equation,

$$x^n + y^n = z^n \quad (n \in \mathbb{N}, n \geq 3),$$

had no solutions for non-zero integers x, y, z. This celebrated assertion, known as Fermat's Last Theorem, has in fact been established as correct owing to the recent work of A. Wiles (1953–). In the 19th century, however, its proof was attempted by many mathematicians among whom was E.E. Kummer (1810–93) who tried heroically to establish a complete result. Arising from his efforts he created a theory of 'ideal numbers' but it fell to Dedekind to introduce 'ideals', albeit in a number-theoretic and so commutative context. For our purposes we shall give a general definition of an 'ideal' appropriate to a possibly non-commutative context.

Definition 10

Let R be a ring and let I be an additive subgroup of R.

1. If for all $x \in R$ and for all $a \in I$, it follows that $xa \in I$, then I is said to be a **left ideal** of R.

2. If for all $x \in R$ and for all $a \in I$, it follows that $ax \in I$, then I is said to be a **right ideal** of R.

3. If for all $x \in R$ and for all $a \in I$, it follows that $xa \in I$ and $ax \in I$, then I is said to be a **two-sided ideal**, or briefly, an **ideal** of R.

R and $\{0\}$ are immediately ideals of R, and $\{0\}$ is called the **zero ideal** of R. If I satisfies 1, 2 or 3 and $I \neq \{0\}$, $I \neq R$, then I is said to be **proper**.

By definition every left or right ideal of R is a subring of R but not every subring of R is necessarily a left or right ideal. Every element of R gives rise to a left or right ideal which may not, however, be proper (see Theorem 11).

Examples 15

1. Let $n \in \mathbb{Z}$. Then $n\mathbb{Z} = \{nx \mid x \in \mathbb{Z}\}$, the subring of those integers divisible by n, is an ideal of $\mathbb{Z}$. (In a commutative ring every left or right ideal is also a two-sided ideal.)

2. Let R be the ring $M_2(\mathbb{Q})$ of all 2×2 matrices over $\mathbb{Q}$. Let

$$L = \left\{ \begin{pmatrix} 0 & p \\ 0 & q \end{pmatrix} \mid p, q \in \mathbb{Q} \right\}.$$

Then L is easily shown to be an additive subgroup of R and L is also a left ideal since, for $x, y, z, t, p, q \in \mathbb{Q}$,

$$\begin{pmatrix} x & y \\ z & t \end{pmatrix} \begin{pmatrix} 0 & p \\ 0 & q \end{pmatrix} = \begin{pmatrix} 0 & xp + yq \\ 0 & zp + tq \end{pmatrix} \in L.$$

Lemma 5

1. Let R be a ring with an identity 1 and let I be a non-zero left ideal of R. If I contains a unit of R then $I = R$.
2. A field has no proper left or right ideals.

Proof

1. Let u be a unit of R such that $u \in I$. Then there exists $v \in R$ such that $vu = 1$. Let $x \in R$. Then $x = x1 = x(va) = (xv)u \in I$ since I is an ideal. Hence $I = R$.
2. Every non-zero element of a field is a unit. Hence the result follows. □

Example 16

We have seen, in the Example above, that $M_2(\mathbb{Q})$ has at least one proper left ideal. We wish to show that $M_2(\mathbb{Q})$ has no proper two-sided ideals. Suppose I is a non-zero ideal of $M_2(\mathbb{Q})$. Let

$$\begin{pmatrix} a & b \\ c & d \end{pmatrix}$$

be a non-zero element of I. Then certainly one of a, b, c, d is non-zero. For the sake of argument we suppose $a \neq 0$. (The reader should consider in turn the cases of non-zero b, c, d and modify the argument to follow.)

Since I is an ideal

$$\begin{pmatrix} 1 & \dfrac{b}{a} \\ 0 & 0 \end{pmatrix} = \begin{pmatrix} \dfrac{1}{a} & 0 \\ 0 & 0 \end{pmatrix} \begin{pmatrix} a & b \\ c & d \end{pmatrix}$$

is an element of I and so also is

$$\begin{pmatrix} 1 & 0 \\ 0 & 0 \end{pmatrix} = \begin{pmatrix} 1 & \frac{b}{a} \\ 0 & 0 \end{pmatrix} \begin{pmatrix} 1 & 0 \\ 0 & 0 \end{pmatrix}.$$

Similarly

$$\begin{pmatrix} 0 & 1 \\ 0 & 0 \end{pmatrix}, \begin{pmatrix} 0 & 0 \\ 1 & 0 \end{pmatrix}, \begin{pmatrix} 0 & 0 \\ 0 & 1 \end{pmatrix}$$

belong to I. Hence, for $x, y, z, t \in \mathbb{Q}$,

$$\begin{pmatrix} x & y \\ z & t \end{pmatrix} = \begin{pmatrix} x & 0 \\ 0 & x \end{pmatrix} \begin{pmatrix} 1 & 0 \\ 0 & 0 \end{pmatrix} + \begin{pmatrix} y & 0 \\ 0 & y \end{pmatrix} \begin{pmatrix} 0 & 1 \\ 0 & 0 \end{pmatrix}$$

$$+ \begin{pmatrix} z & 0 \\ 0 & z \end{pmatrix} \begin{pmatrix} 0 & 0 \\ 1 & 0 \end{pmatrix} + \begin{pmatrix} t & 0 \\ 0 & t \end{pmatrix} \begin{pmatrix} 0 & 0 \\ 0 & 1 \end{pmatrix} \in I.$$

Thus $I = M_2(\mathbb{Q})$.

The next theorem, which may be passed over on a first reading, brings together, nevertheless, some useful facts and, at the same time, serves to define some convenient notation.

Theorem 11

Let R be a ring.

1. Let $a \in R$. Then Ra defined by

$$Ra = \{xa \,|\, x \in R\}$$

   is a left ideal of R.

2. Let $L_1, L_2, \ldots, L_n$ be left ideals of R. Then $L_1 + L_2 + \ldots + L_n$ defined by

$$L_1 + L_2 + \ldots + L_n = \{a_1 + a_2 + \ldots + a_n \,|\, a_i \in L_i, \, i = 1, 2, \ldots, n\}$$

   is a left ideal of R.

3. Let $a_1, a_2, \ldots, a_n \in R$. Then

$$Ra_1 + Ra_2 + \ldots + Ra_n = \{x_1 a_1 + x_2 a_2 + \ldots + x_n a_n \,|\, x_i \in R, \, i = 1, 2, \ldots, n\}$$

   is a left ideal of R.

Proof

1. $0 = 0a \in Ra$, and for all $x_1, x_2, y \in R$ we have

$$x_1 a + x_2 a = (x_1 + x_2)a \in Ra, \quad -(x_1 a) = (-x_1)a \in Ra.$$

   Thus Ra is a left ideal.

2. We prove the result for two left ideals A and B. The general result is obtained by a simple induction. $A + B = \{a + b \,|\, a \in A, b \in B\}$ and $A + B$ is an additive subgroup of R. Let $x \in R$, $a \in A$, $b \in B$. Then we have $x(a + b) = xa + xb \in A + B$ since A, B are left ideals. Thus $A + B$ is a left ideal.

3. This follows from 1 and 2 above. $\qquad\qquad\qquad\qquad\qquad\qquad\square$

In our considerations of groups we found that normal subgroups of a group and the homomorphisms of a group were important and directly related concepts. A similar relationship will be seen to exist between ideals of a ring and 'homomorphisms' of a ring.

Definition 11

Let R and S be rings and let $f : R \to S$ be a mapping such that

$$f(x + y) = f(x) + f(y), \; f(xy) = f(x)f(y) \quad \text{(for all } x, y \in R\text{)}.$$

Then f is said to be a **homomorphism** from R to S.

The homomorphism f, as defined above, preserves the additive and multiplicative structure of R in its image $f(R)$ in T. Thus f is a homomorphism of R considered as an additive Abelian group and considered as a multiplicative semigroup.

Examples 17

1. Let n be a given integer, $n > 0$. Then the mapping $f : \mathbb{Z} \to \mathbb{Z}_n$ given by

$$f(a) = \bar{a}$$

   where $\bar{a}$, the equivalence class to which a belongs $(a \in \mathbb{Z})$, is a homomorphism since we know that

$$\overline{a + b} = \bar{a} + \bar{b},$$

$$\overline{ab} = \bar{a}\bar{b}.$$

2. Let $F[x]$ be a polynomial ring over a field F and let $\alpha \in F$. Then let us define $f : F[x] \to F$ by

$$f : a(x) \to a(\alpha) \quad (a(x) \in F[x]).$$

In other words, f replaces a polynomial by the result of substituting α for x in the polynomial. But we know that

$$a(x) + b(x) = c(x),\ a(x)b(x) = d(x) \quad (a(x), b(x), c(x), d(x) \in F[x])$$

implies that

$$a(\alpha) + b(\alpha) = c(\alpha),\ a(\alpha)b(\alpha) = d(\alpha)$$

and so f is a homomorphism.

Theorem 12

Let R and S be rings and let $f : R \to S$ be a homomorphism. Then the following statements hold.

1. If O_R and O_S are the zero elements of R and S respectively then $f(O_R) = O_S$.
2. For all $x \in R$, $-f(x) = f(-x)$ where $-f(x)$ is the inverse with respect to addition of $f(x)$ in S and $-x$ is the inverse with respect to addition of x in R.
3. $f(R)$ is a subring of S.
4. Let $K = \{x \in R | f(x) = O_S\}$, O_S defined in 1. Then K is an ideal of R.

Proof

1 and 2 hold since f is a homomorphism of R into S where both rings are considered solely as additive groups. Furthermore, by the same reasoning, $f(R)$ is an additive subgroup of S. To prove that $f(R)$ is indeed a subring of S we have merely to establish multiplicative closure of $f(R)$ in S and this occurs since f is a homomorphism from R to S considered as multiplicative semigroups. Hence, finally, we conclude that $f(R)$ is a subring of S.

To prove 4, we observe that K is the kernel of the homomorphism f from the additive group R to the additive group S. Let now $a \in K$, $x \in R$. Then

$$f(xa) = f(x)f(a) = f(x)O_S = O_S,$$
$$f(ax) = f(a)f(x) = O_S f(x) = O_S.$$

Hence $xa, ax \in K$ and so K is an ideal.   □

Definition 12

Let R and S be rings and let $f : R \to S$ be a homomorphism. Then $K = \{x \in R | f(x) = O_S\}$ is an ideal of R which is called the **kernel** of f, written Ker f.

We now come to several definitions and results which correspond to definitions and results previously considered in Chapter 4. As the results and proofs are so similar, those to be provided here will be less detailed than usual. The reader should, however, ensure that he or she fully understands the arguments.

We recall that an ideal I of a ring R is, in fact, an additive subgroup of the additive group R and so we have cosets of I in R where the coset containing $x \in R$ is

$$x + I = \{x + a \,|\, a \in I\}.$$

Lemma 6

Let R and S be rings and let $f : R \to S$ be a homomorphism with kernel K. Let x and y be elements of R. Then $f(x) = f(y)$ if and only if $x + K = y + K$.

Proof

We have shown previously that $f(x) = f(y)$ if and only if the cosets $x + K$ and $y + K$ are equal. $\square$

Theorem 13

Let I be an ideal of the ring R. Let R/I denote the set of cosets of I in R. Two binary operations are defined on R/I as follows:

Let $x + I$ and $y + I$ be cosets of I in R and define

$$(x + I) + (y + I) = (x + y) + I,$$
$$(x + I)(y + I) = xyI.$$

Then the binary operations are well-defined and under these operations R/I is a ring. Further the mapping $R \to R/I$ given by $x \to x + I$ $(x \in R)$ is a homomorphism of R onto R/I with kernel I.

Proof

Certainly under the definition of addition R/I is an additive group. We have to show that multiplication is well-defined in R/I. Thus we have to show that if $x + I = x' + I$ $(x, x' \in R)$ and $y + I = y' + I$ $(y, y' \in R)$ then $xy + I = x'y' + I$. But certainly $x + I = x' + I$ implies that $x' = x + a$ for some $a \in I$, also $y' = y + b$ for some $b \in I$. Then $x'y' = (x + a)(y + b) = xy + ay + xb + ab$. But I is an ideal and hence we have $ay + xb + ab \in I$. Thus $x'y' + I = xy + (ay + xb + ab) + I = xy + I$. Our definition of multiplication is well founded.

To prove the associativity of multiplication we have for all $x, y, z \in R$

$$[(x+I)(y+I)](z+I) = (xy+I)(z+I)$$
$$= (xy)z + I$$
$$= x(yz) + I$$
$$= (x+I)(yz+I)$$
$$= (x+I)[(y+I)(z+I)].$$

The distribution laws may be proved and we conclude that R/I is a ring.

The mapping $f : R \to R/I$ given by $f(x) = x + I$ $(x \in R)$ is a homomorphism of the additive group R onto R/I. For $x, y \in R$ we have

$$f(xy) = xy + I = (x+I)(y+I) = f(x)f(y)$$

and so f is a ring homomorphism. Finally

$$\text{Ker } f = \{x \in R | f(x) = I\} = \{x \in R | x + I = I\} = I. \qquad \square$$

Definition 13

Let I be an ideal of the ring R. The ring R/I, as considered above, is said to be a **factor-ring** of R.

Example 18

For each $n \in \mathbb{Z}$, $n\mathbb{Z} = \{nx \mid x \in \mathbb{Z}\}$ is an ideal of $\mathbb{Z}$ and the factor-ring $\mathbb{Z}/n\mathbb{Z}$ is the ring which we have called $\mathbb{Z}_n$.

Definition 14

Let R and S be rings and let $f : R \to S$ be a homomorphism. If f is surjective then f is said to be an **epimorphism**. If f is injective then f is said to be a **monomorphism**. If f is bijective then f is said to be an **isomorphism**.

Remark

In Theorem 8 we had an 'embedding', as it was called, of an integral domain D into its quotient field F. By an embedding we now understand an isomorphism and we have therefore already shown that there exists an isomorphic copy of D contained in F. In this instance we may identify D with its isomorphic copy and so regard D as a sub-integral domain of F, just as we regard $\mathbb{Z}$ as embedded in, and a sub-integral domain of, $\mathbb{Q}$.

Theorem 14 First Isomorphism Theorem

Let R and S be rings and let $f : R \to S$ be an epimorphism with kernel K. Then R/K and S are isomorphic.

Proof

K is an ideal of R and so we have an epimorphism $R \to R/K$. We now define $h : R/K \to S$ by

$$h(x + K) = f(x) \quad (x \in R).$$

By the proof of the First Isomorphism Theorem for groups h is a well-defined mapping which is, in fact, an isomorphism between the additive groups R/K and S. We merely have to show that h preserves multiplication. Let x, $y \in R$, then we have $h[(x + K)(y + K)] = h[xy + K] = f(xy) = f(x)f(y) = h(x + K)h(y + K)$. The proof is finally complete. $\qquad \Box$

Lemma 7

Let R be a ring. Let S be a subring of R and let I be an ideal of R. Then $S + I = \{x + a \mid x \in S, a \in I\}$ is a subring of R and $I \cap S$ is an ideal of S.

Proof

We leave this proof to the Exercises. $\qquad \Box$

Theorem 15 Second Isomorphism Theorem

Let R be a ring. Let S be a subring of R and let I be an ideal of R. Then the factor-rings $(S + I)/I$ and $S/(I \cap S)$ are isomorphic.

Proof

We define $p : S \to (S + I)/I$ by

$$p(x) = x + I \quad (x \in S).$$

Then p is an epimorphism. If K is the kernel of p then, by the First Isomorphism Theorem, S/K is isomorphic to $(S + I)/I$.

On proving that $K = I \cap S$ the proof is complete. $\qquad \Box$

Exercises 5.4

1. Prove that $A = \left\{ \begin{pmatrix} x & y \\ 0 & 0 \end{pmatrix} | x, t \in Q \right\}$ is a right ideal of $M_2(\mathbb{Q})$.

2. Let L_1 and L_2 be left ideals of a ring R. Prove that $L_1 \cap L_2$ is a left ideal of R. Generalize.

3. Let $L_1 \subseteq L_2 \subseteq L_3 \subseteq \dots$ be a countable ascending chain of left ideals of a ring R. Prove that $\bigcup_{i=1}^{\infty} L_i$ is a left ideal of R.

4. Let X be a non-empty subset of a ring R. Let

$$A = \{a \in R | xa = 0 \text{ for all } x \in X\}.$$

Prove that A is a right ideal of R.

5. Let R be a ring not necessarily possessing an identity. Let $a \in R$. Prove that $\{na + xa | n \in \mathbb{Z}, x \in R\}$ is a left ideal of R containing a.

6. Let R and S be rings and let $f : R \to S$ be a homomorphism. Let L be a left ideal of S. Prove that $\{x \in R | f(x) \in L\}$ is a left ideal of R.

7. Let R and S be rings and let $f : R \to S$ be an epimorphism. Let L be a left ideal of R. Prove that $f(L)$ is a left ideal of S.

8. Let R be a ring and let S be a subring of R. Let I be an ideal of R. Prove that $S + I = \{x + a | x \in S, a \in I\}$ is a subring of R and $I \cap S$ is an ideal of S.

9. Let R, S and T be rings and let $f : R \to S$ and $g : S \to T$ be given homomorphisms. Prove that $g \circ f$ is a homomorphism from R to T.

10. Let R be the ring of 2×2 matrices of the form

$$\begin{pmatrix} a & b \\ -b & a \end{pmatrix} \quad (a, b \in R).$$

Prove that the mapping

$$a + ib \to \begin{pmatrix} a & b \\ -b & a \end{pmatrix} \quad (a, b \in R)$$

is an isomorphism of $\mathbb{C}$ with R.

5.5 Principal Ideal and Unique Factorization Domains

We here continue our study of integral domains with a view to obtaining properties very similar to those of the integral domain of the integers $\mathbb{Z}$.

Definition 15

Let D be an integral domain. Let $a, b \in D$, $b \neq 0$. Then b is said to be a **divisor** of a if there exists $c \in D$ such that $a = bc$.

We recall that u is a unit of a ring R if there exists $v \in R$ such that $uv = vu = 1$.

Definition 16

Let D be an integral domain. Let $a, b \in D$. Then a is an **associate** of b if there exists a unit u in D such that $a = bu$.

Notice that if a is an associate of b then b is an associate of a for if $a = bu$ where $uv = vu = 1$ then $av = buv = b1 = b$. An element u is a unit if and only if u is an associate of 1.

We further recall that if R is a ring and $a \in R$ then $Ra = \{xa | a \in R\}$ and if, in particular, R has an identity 1 then $a = 1a \in Ra$.

Lemma 8

Let D be an integral domain. Let $a, b \in D$. Then b divides a if and only if $Da \subseteq Db$.

Proof

If b divides a then $a = cb$ for some $c \in D$. Then $Da = Dcb \subseteq Db$. Conversely if $Da \subseteq Db$ then $a \in Da \subseteq Db$ and so there exists $c \in D$ such that $a = cb$. $\square$

Lemma 9

Let D be an integral domain. Let $a, b \in D$. The following statements are equivalent.

1. $Da = Db$.
2. a divides b and b divides a.
3. a and b are associates.
4. $a = bu$ for some unit u of D.

Proof

By Lemma 8, 1 and 2 are equivalent and by remarks above 3 and 4 are equivalent. We prove the equivalence of 2 and 3.

If a divides b and b divides a there exist $c, d \in D$ such that $b = ac$, $a = bd$. Then $a = bd = acd$ from which, as D is an integral domain, $1 = cd$ and so d and c are units. Hence a and b are associates. On the other hand if a and b are associates there exist units $u, v \in D$ such that $a = bu$ and $b = av$. Hence b divides a and a divides b. This completes the proof.   □

Examples 19

1. Two integers m, n are associates in $\mathbb{Z}$ if and only if $m = \pm n$. In $\mathbb{Z}$ Lemma 9 is obviously true.
2. Let F be a field. The units of $F[x]$, the polynomial ring of F over x, are the non-zero elements of F. Therefore $f(x), g(x) \in F[x]$ are associates if and only if $f(x) = cg(x)$ where $c \in F$, $c \neq 0$.

Definition 17

Let D be an integral domain. A non-zero element p of D, which is not a unit, is said to be **irreducible** if whenever $a \in D$ and a divides p then a is an associate of 1 or p.

Definition 18

Let D be an integral domain. A non-zero element p of D, which is not a unit, is said to be a **prime** if whenever p divides ab $(a, b \in D)$ then p divides a or p divides b.

For our convenience we make the following definition for a restricted class of rings; we content ourselves with remarking that, with appropriate modification, the restrictions in the definition may be removed.

Definition 19

Let R be a commutative ring with an identity.

1. An ideal P of R, $P \neq R$, is said to be **prime** if whenever $a, b \in R$ and $ab \in P$, then either $a \in P$ or $b \in P$.
2. An ideal M of R, $M \neq R$, is said to be **maximal** if whenever $M \subseteq I$ and $I \subseteq R$ for any ideal I, then either $M = I$ or $I = R$.

Theorem 16

Let R be a commutative ring with an identity.

1. An ideal P of R is prime if and only if R/P is an integral domain.
2. An ideal M of R is maximal if and only if R/M is a field.

Proof

We note that the factor-ring in 1 and 2 is, at least, a commutative ring with an identity 1.

1. Let P be prime. Let $a, b \in R$ be such that $(a + P)(b + P) = P$. Then we have $ab + P = P$ and so $ab \in P$. As P is prime $a \in P$ or $b \in P$ which implies $a + P = P$ or $b + P = P$. Hence, R/P has no proper divisors of zero and so is an integral domain.

   Let now R/P be an integral domain. Let $a, b \in R$ be such that $ab \in P$. Then $(a + P)(b + P) = ab + P = P$ and so we have divisors of zero in R/P. Since R/P is an integral domain $a + P = P$ or $b + P = P$ from which $a \in P$ or $b \in P$. Hence P is prime.

2. Let M be maximal. Let $a \in R$ be such that $a + M \neq M$. Then $a \notin M$ and so the ideal $Ra + M$ contains a and M and so $M \subset Ra + M$. Since M is a maximal ideal $Ra + M = R$. Then there exist $x \in R$ and $m \in M$ such that

$$xa + m = 1.$$

Hence

$$1 + M = (xa + m) + M = xa + m + M = xa + M = (x + M)(a + M).$$

Thus $x + M$ is the inverse of $a + M$. Hence R/M is a field.

   Let now R/M be a field. Let I be an ideal such that $M \subseteq I \subseteq R$. If $M \neq I$ there exists $a \in I$, $a \notin M$. Then $a + M \neq M$ and so, as R/M is a field, $a + M$ is invertible in R/M and there exists $b \in R$ such that $(a + M)(b + M) = 1 + M$. Therefore $ab = 1 + m$ for some $m \in M$. Hence, as $a \in I$, $m \in M$ and $M \subseteq I$ we have $1 \in I$. But then $I = R$. $\qquad \square$

Theorem 17

Let D be an integral domain. Then a prime is also an irreducible element.

Proof

Let p be a prime in D. Let $a \in D$ and suppose a divides p. We wish to show that a is an associate of 1 or p. We have $p = ab$ for some $b \in D$. Since p is prime p divides ab and so either p divides a or p divides b. If p divides a then $a = pc$ for some $c \in D$ and so $p = ab = pcb$ and $1 = cb$. Thus b is a unit and p and a are associates. On the other hand, if p divides b then $b = dp$ for some $d \in D$ and so we have $p = ab = adp$ and $1 = ad$. Thus a is an associate of 1. This completes the proof. $\qquad\square$

In the integral domain $\mathbb{Z}$ every irreducible element is also prime. In analogy with $\mathbb{Z}$ we wish to investigate those integral domains in which every irreducible element is prime. To facilitate our investigation it is convenient to make a definition which is obviously modelled on the Fundamental Theorem of Arithmetic in $\mathbb{Z}$.

Definition 20

Let D be an integral domain in which every non-zero element, which is not a unit, is expressible as a finite product of irreducible elements. Furthermore, whenever a non-zero element x of D which is not a unit, is written as

$$x = p_1 p_2 \ldots p_m = q_1 q_2 \ldots q_n$$

where $p_1, p_2, \ldots, p_m; q_1, q_2, \ldots, q_n$ are irreducible elements then $m = n$ and, with a suitable reordering if necessary, p_i and q_i are associates $(i = 1, 2, \ldots, n)$. Then D is said to be a **unique factorization domain** (U.F.D.).

By the Fundamental Theorem of Arithmetic $\mathbb{Z}$ is a U.F.D.

Theorem 18

In a unique factorization domain every irreducible element is prime.

Proof

Let D be a U.F.D. Let x be an irreducible element of D. Suppose x divides ab $(a, b \in D)$. We have to show that x divides a or x divides b. By supposition there exists $y \in D$ such that

$$xy = ab.$$

Now y, a and b are products of irreducible elements, say,

$$a = p_1 p_2 \ldots p_m, \, b = q_1 q_2 \ldots q_n, \, y = r_1 r_2 \ldots r_t,$$

where $p_1, p_2, \ldots, p_m; \, q_1, q_2, \ldots, q_n; \, r_1, r_2, \ldots, r_t$ are irreducible elements. Then

$$x r_1 r_2 \ldots r_t = p_1 p_2 \ldots p_m q_1 q_2 \ldots q_n.$$

But these are two factorizations into irreducible elements and so, as D is a U.F.D., we must have $1 + t = m + n$ and, more importantly, every irreducible element on the left-hand side must be an associate of an irreducible element on the right-hand side. Thus x is an associate of some p_i or some q_j. But this implies that x divides a or x divides b. $\qquad\square$

Definition 21

Let D be an integral domain. Let a, b, c be non-zero elements of D. Then c is said to be a **common divisor** of a and b if c divides a and c divides b. A common divisor d, which is itself divisible by any other common divisor, is called a **greatest common divisor** (G.C.D.).

The terminology of this definition corresponds to the terminology of Definitions 3 and 4 of Chapter 2 in the case of the integers. As in that case, we may show that in a U.F.D. any two non-zero elements have a G.C.D. We defer the discussion of a G.C.D. until after we have discussed 'principal ideal rings'.

Definition 22

Let D be an integral domain. Let $a \in D$. Then $Da = \{xa \mid x \in D\}$ is said to be a **principal** ideal. An integral domain in which every ideal is principal is called a **principal ideal domain** (P.I.D.).

Lemma 10

Let D be a principal ideal domain. Any two non-zero elements a and b of D have a G.C.D. d given by

$$Da + Db = Dd.$$

Proof

Da and Db are ideals of D and so $Da + Db$ is also an ideal. Hence as D is a P.I.D. there exists $d \in D$ such that $Da + Db = Dd$. We claim that d is indeed a G.C.D. of a and b. Certainly $Da \subseteq Dd$ and so d divides a and similarly d divides b. Thus

d is a common divisor of a and b. Suppose $c \in D$ is also a common divisor of a and b. Now $d \in Da + Db$ and so there exist $x, y \in D$ such that $d = xa + yb$. But now if c divides a and c divides b we must have that c divides d. Thus d is a G.C.D. of a and b. $\square$

We now relate our earlier discussion of Euclidean domains to principal ideal domains.

Theorem 19

A Euclidean domain is a principal ideal domain.

Proof

Let D be a Euclidean domain with mapping $v : D \setminus \{0\} \to \{0, 1, 2, \}$. Let I be an ideal of D. We wish to prove that I is a principal ideal. If $I = \{0\}$ the result is true so suppose $I \neq \{0\}$. Choose an element b of I for which, amongst all the non-zero elements of I, $v(b)$ is as small as possible. As I is an ideal $Db \subseteq I$. We shall show that $Db = I$. Let $a \in I$. As D is a Euclidean domain there exist $q, r \in D$ such that

$$a = bq + r$$

where either $r = 0$ or, if $r \neq 0$, then $v(r) < v(b)$. But, as I is an ideal,

$$r = a - bq \in I.$$

But if $r \neq 0$ we have $v(r) < v(b)$ and this contradicts the choice of b as being such that $v(b)$ is as small as possible. Thus $r = 0$ and $a = bq \in Db$. Hence, finally, $Db = I$. $\square$

Examples 20

1. $\mathbb{Z}$ is a Euclidean domain and we now know that every ideal of $\mathbb{Z}$ is of the form $n\mathbb{Z} = \{nx \mid x \in \mathbb{Z}\}$ for some n.
2. The polynomial ring $F[x]$, where F is a field, is a Euclidean domain and so is a P.I.D.

Theorem 20

In a principal ideal domain every irreducible element is prime.

Proof

Let D be a P.I.D. Let p be an irreducible element of D. Let p divide ab where $a, b \in D$. We wish to show that p divides a or p divides b.

Suppose p does not divide a. Let $c \in D$ be such that c divides p and c divides a. Since p is irreducible c is a unit or an associate of p. If c is an associate of p then as c divides a so also does p divide a which is false. Hence c is a unit. Hence, by Lemma 10,

$$Da + Dp = Dd$$

where d is necessarily a unit and so $Dd = D$, giving $Da + Dp = D$. Hence there exist $x, y \in D$ such that

$$xa + yp = 1.$$

But then

$$xab + ypb = b$$

from which p divides b. Hence p is a prime.   □

Corollary

Let D be a principal ideal domain. Let p be an irreducible element of D. Let $a_1, a_2, \ldots, a_n \in D$ and suppose p divides $a_1 a_2 \ldots a_n$. Then for some i $(1 \leq i \leq n)$, p divides a_i.

Proof

Either p divides a_1 or p divides $a_2 a_3 \ldots a_n$. A simple induction proves the result.   □

Our aim is to prove that a P.I.D. is necessarily a U.F.D. We are thereby enabled to deduce that a Euclidean domain is a U.F.D. At a first reading the reader may wish to confine himself or herself to the fact of the deduction and to avoid the necessary proofs. We have a lengthy lemma before the main theorem.

Lemma 11

Let D be a principal ideal domain. Let a be a non-zero element of D which is not a unit of D. Then a is the product of a finite number of irreducible elements.

Proof

We argue by contradiction. Suppose therefore that a is not the finite product of irreducible elements.

Since a cannot itself be irreducible, a is divisible by a_1, say, $a_1 \in D$, where a_1 is neither a unit nor an associate of a. Hence there exists $b_1 \in D$ which is also neither a unit nor an associate of a, such that

$$a = a_1 b_1.$$

But this implies that

$$Da \subseteq Da_1$$

and since a_1 is not an associate of a we have

$$Da \subset Da_1.$$

Now either a_1 or b_1 is not a finite product of irreducible elements. Suppose a_1 is not a product of irreducible elements. We apply the same argument to a_1 which is not a product of irreducible elements and we obtain a_2 dividing a_1 where

$$Da_1 \subset Da_2$$

and where a_2 is not a product of irreducible elements. We continue this process to obtain a strictly ascending chain of ideals Da_n $(n \in \mathbb{N})$ such that

$$Da \subset Da_1 \subset Da_2 \subset \ldots$$

where each a_n $(n \in \mathbb{N})$ is not a product of irreducible elements. Let

$$I = \bigcup_{n=1}^{\infty} Da_n.$$

Then I is an ideal of D (see Exercises 5.4, no. 3) and so $I = Dc$ for some $c \in D$. But c must belong to some subset of the infinite union and so there must exist k such that $c \in Dc_k$. But then

$$I = Dc \subseteq Da_k \subseteq I$$

and so $I = Da_k$. But this implies that $Da_k = Da_{k+1} = \ldots$ and this contradicts the fact that the ideals in the chain are distinct. Hence our original supposition is false and a is indeed a finite product of irreducible elements. $\qquad \square$

Theorem 21

A principal ideal domain is also a unique factorization domain.

Proof

Let D be a P.I.D. By Lemma 11 we have shown that every non-zero element of D which is not a unit is a finite product of irreducible elements. We have to show that such a product is essentially unique. Let $a \in D$ and let a be expressed as

$$a = p_1 p_2 \cdots p_m = q_1 q_2 \cdots q_n$$

where $p_1, p_2, \ldots, p_m; q_1, q_2, \ldots, q_n$ are irreducible elements of D. Then p_1 divides $q_1 q_2 \cdots q_n$ and so, by the Corollary above, p_1 divides one of $q_1, q_2, \ldots, q_n$. By renumbering, if necessary, we may suppose p_1 divides q_1. Then p_1 and q_1 are irreducible elements which are associates and so $q_1 = u p_1$ where u is a unit of D.

Then

$$p_1 p_2 \cdots p_m = q_1 q_2 \cdots q_n = u p_1 q_2 q_3 \cdots q_n$$

which implies that

$$p_2 p_3 \cdots p_m = u q_2 q_3 \cdots q_n = q_2' q_3 \cdots q_n$$

where $q_2' = u q_2$ is an irreducible element. A simple induction now proves the desired result. □

Corollary

A Euclidean domain is also a unique factorization domain.

Exercises 5.5

1. Let R be a commutative ring with an identity. Prove that a maximal ideal of R is a prime ideal. Give an example of a ring R and a prime ideal of R which is not maximal in R.

2. $\mathbb{Z}[i]$ is the ring of Gaussian integers. Find the G.C.D. d in $\mathbb{Z}[i]$ of $4 + 2i$ and $1 - 3i$ and write d in the form

$$(4 + 2i)z_1 + (1 - 3i)z_2 = d$$

for appropriate $z_1, z_2 \in \mathbb{Z}[i]$.

3. Let D be a principal ideal domain. Prove that any n non-zero elements $a_1, a_2, \ldots, a_n$ of D have a G.C.D. d given by

$$D a_1 + D a_2 + \ldots + D a_n = D d.$$

5.6 Factorization in $\mathbb{Q}[x]$

Given a non-zero polynomial $f(x) \in \mathbb{Q}[x]$ we want to devise a criterion by which the polynomial is a prime, or is equivalently irreducible, in $\mathbb{Q}[x]$. We remark first that any such polynomial $f(x)$ may be written as

$$f(x) = \frac{1}{c} g(x)$$

where $g(x) \in \mathbb{Z}[x]$ and $c \in \mathbb{Z}$.

Example 21

$$f(x) = \frac{2}{3} x^2 + \frac{1}{4} x + \frac{1}{2} = \frac{1}{12} \left(8x^2 + 3x + 6\right) = \frac{1}{12} g(x).$$

It suffices therefore to consider polynomials which are in $\mathbb{Z}[x]$ but to attempt their factorization in $\mathbb{Q}[x]$. We aim to show that if a polynomial in $\mathbb{Z}[x]$ factors non-trivially in $\mathbb{Q}[x]$ then it already has a non-trivial factorization in $\mathbb{Z}[x]$. From this fact we infer that if a polynomial in $\mathbb{Z}[x]$ does not factor non-trivially in $\mathbb{Z}[x]$ then it does not factor non-trivially in $\mathbb{Q}[x]$.

Definition 23

Let $f(x)$ be a polynomial in $\mathbb{Z}[x]$. Then the greatest common divisor of the coefficients of $f(x)$ is called the **content** of $f(x)$. A polynomial of content 1 is said to be **primitive**.

Let $f(x) \in \mathbb{Q}[x]$, $f(x) \neq 0$. Then we may write

$$f(x) = \frac{d}{c} h(x)$$

where c, d are integers, $h(x) \in \mathbb{Z}[x]$ and $h(x)$ has content 1.

We illustrate this assertion by means of Examples.

Examples 22

1. $f(x) = \dfrac{14}{5} x^2 + \dfrac{28}{3} x + \dfrac{21}{2} = \dfrac{1}{2.3.5} [84x^2 + 280x + 315]$

$$= \frac{7}{30} [12x^2 + 40x + 45] = \frac{7}{30} h(x),$$

where $h(x)$ has content 1.

2. $f(x) = \dfrac{5}{12} x^3 + \dfrac{1}{8} x^2 + \dfrac{11}{6} x + \dfrac{1}{14}$

$\qquad = \dfrac{1}{8.3.7} [70x^3 + 21x^2 + 308x + 12]$

$\qquad = \dfrac{1}{168} h(x),$

where $h(x)$ has content 1.

Lemma 12

Let $f(x)$ and $g(x)$ be primitive polynomials in $\mathbb{Z}[x]$. Suppose there exist $c_1, c_2 \in \mathbb{Z}$, $c_1 \neq 0$, $c_2 \neq 0$, such that $c_1 f(x) = c_2 g(x)$. Then $c_1 = \pm c_2$ and $f(x) = \pm g(x)$.

Proof

Let $f(x) = a_0 + a_1 x + \ldots + a_n x^n$ $(a_n \neq 0)$. Then the G.C.D. of $a_0, a_1, \ldots, a_n$ is 1 and so there exist $t_0, t_1, \ldots, t_n \in \mathbb{Z}$ such that

$$t_0 a_0 + t_1 a_1 + \ldots + t_n a_n = 1.$$

Since $c_1 f(x) = c_2 g(x)$, c_2 divides $c_1 a_0, c_1 a_1, \ldots, c_1 a_n$ and so c_2 divides

$$t_0 c_1 a_0 + t_1 c_1 a_1 + \ldots + t_n c_n a_n = c_1 (t_0 a_0 + t_1 a_1 + \ldots + t_n a_n) = c_1.$$

Similarly c_1 divides c_2. Thus $c_1 = \pm c_2$ and so $f(x) = \pm g(x)$. □

The next lemma holds the key to the use of primitive polynomials.

Lemma 13 Gauss's Lemma

Let $f(x)$ and $g(x)$ be primitive polynomials in $\mathbb{Z}[x]$. Then $f(x)g(x)$ is primitive.

Proof

Let
$$f(x) = a_0 + a_1 x + \ldots + a_m x^m \quad (a_m \neq 0),$$
$$g(x) = b_0 + b_1 x + \ldots + b_n x^n \quad (b_n \neq 0),$$
$$f(x)g(x) = h(x) = c_0 + c_1 x + \ldots + c_{m+n} x^{m+n} \quad (c_{m+n} \neq 0).$$

If $h(x)$ is not primitive there exists a prime $p \in \mathbb{Z}$ such that p divides each of $c_0, c_1, \ldots, c_{m+n}$. Now p cannot divide all of the coefficients of $f(x)$ or all of the coefficients of $g(x)$ since $f(x)$ and $g(x)$ are primitive. Suppose therefore that p divides $a_0, a_1, \ldots, a_{r-1}$ but p does not divide a_r where $0 \le r < m$ and that p divides $b_0, b_1, \ldots, b_{s-1}$ but p does not divide b_s where $0 \le s < n$. Consider c_{r+s}. We have

$$c_{r+s} = a_0 b_{r+s} + a_1 b_{r+s-1} + \ldots + a_{r-1} b_{s+1} + a_r b_s + a_{r+1} b_{s-1} + \ldots + a_{r+s} b_0.$$

Now p divides c_{r+s}; $a_0, a_1, , a_{r-1}$; $b_0, b_1, \ldots, b_{s-1}$ and hence it follows that p divides $a_r b_s$ and so divides a_r or b_s. But this is a contradiction and so $f(x)g(x)$ is primitive.   □

Example 23

$2x^2 + 3x + 1, x^2 + 2$ are primitive and so is

$$(2x^2 + 3x + 1)(x^2 + 2) = 2x^4 + 3x^3 + 5x^2 + 6x + 2.$$

We use Lemma 13 to show that if factorization does not occur in $\mathbb{Z}[x]$ then it does not occur in $\mathbb{Q}[x]$.

Lemma 14

Let $f(x) \in \mathbb{Z}[x]$ and let $f(x)$ have degree n, $n > 0$. Suppose that $f(x)$ does not factorize in $\mathbb{Z}[x]$ into the product of two polynomials of degrees r and s where $0 < r < n$, $0 < s < n$ $(r + s = n)$. Then $f(x)$ does not factorize in $\mathbb{Q}[x]$ into the product of two polynomials of degrees r and s.

Proof

We argue by contradiction and suppose

$$f(x) = g(x)h(x)$$

where $g(x)$ and $h(x)$ are polynomials in $\mathbb{Q}[x]$ of degrees r and s respectively. Now

$$f(x) = c_0 f_0(x), \quad g(x) = \frac{c_1}{d_1} g_0(x), \quad h(x) = \frac{c_2}{d_2} h_0(x),$$

where $f_0(x)$, $g_0(x)$ and $h_0(x)$ are primitive polynomials in $\mathbb{Z}[x]$ and c_0, c_1, c_2, d_1, d_2 are in $\mathbb{Z}$. Then

$$c_0 f_0(x) = \frac{c_1}{d_1} g_0(x) \frac{c_2}{d_2} h_0(x)$$

and so

$$c_0 d_1 d_2 f_0(x) = c_1 c_2 g_0(x) h_0(x).$$

Now, by Gauss's Lemma, $g_0(x)h_0(x)$ is primitive and so, by Lemma 12, $c_0d_1d_2 = \pm c_1c_2$ and $f_0(x) = \pm g_0(x)h_0(x)$. But $g_0(x)$ and $h_0(x)$ have degrees r and s respectively and so we have derived a contradiction since $f(x) = c_0 f_0(x) = \pm g_0(x)h_0(x)$. Thus the lemma is proved. $\qquad\square$

We now obtain our main result which is named after F.G.M. Eisenstein (1823–52).

Theorem 22 Eisenstein's Criterion

Let $f(x) \in \mathbb{Z}[x]$ and let $f(x) = a_0 + a_1x + \ldots + a_nx^n$ $(a_n \neq 0)$. Suppose there exists a prime p such that:

1. p divides $a_0, a_1, \ldots, a_{n-1}$,
2. p does not divide a_n and
3. p^2 does not divide a_0.

Then $f(x)$ is irreducible as a polynomial in $\mathbb{Q}[x]$.

Proof

We argue by contradiction. If $f(x)$ is not irreducible in $\mathbb{Q}[x]$ then $f(x)$ is not prime in $\mathbb{Q}[x]$ and so $f(x)$ may be factorized in $\mathbb{Q}[x]$ into two polynomials of degrees r and s where $0 < r < n$, $0 < s < n$, $r + s = n$. There is necessarily a corresponding factorization of $f(x)$ in $\mathbb{Z}[x]$. Hence we may suppose that

$$f(x) = g(x)h(x)$$

where $g(x)$ and $h(x)$ are polynomials in $\mathbb{Z}[x]$ of degrees r and s respectively. Let

$$g(x) = b_0 + b_1x + \ldots + b_rx^r \quad (b_r \neq 0),$$
$$h(x) = c_0 + c_1x + \ldots + c_sx^s \quad (c_s \neq 0).$$

Now $a_0 = b_0c_0$ and since p divides a_0 but p^2 does not divide a_0, either b_0 or c_0, but not both b_0 and c_0, is divisible by p. Suppose p divides b_0 but p does not divide c_0. If p were to divide $b_0, b_1, \ldots, b_r$ then all the coefficients of $f(x)$ would also be divisible by p and that is false. We suppose therefore that p divides $b_0, b_1, \ldots, b_{k-1}$ but p does not divide b_k for some k where $0 < k \leq r < n$. Since

$$a_k = b_kc_0 + b_{k-1}c_1 + \ldots + b_0c_k$$

we have that p divides a_k; $b_0, b_1, \ldots, b_{k-1}$ and so p divides b_kc_0. But p does not divide c_0 and so p divides b_k which is false. Hence our initial assumption was wrong and consequently $f(x)$ is irreducible in $\mathbb{Q}[x]$. $\qquad\square$

We give some examples of the use of Eisenstein's Criterion.

Examples 24

1. Let p be a prime. Then $x^n - p \in \mathbb{Q}[x]$ is irreducible by the criterion.
2. $21 + 6x^2 + 9x^3 + 4x^4$ is irreducible in $\mathbb{Q}[x]$ on applying the criterion with $p = 3$.
3. We may also prove that $f(x) = 1 + x + x^2 + \ldots + x^{p-1}$ is irreducible where p is a prime. We cannot apply the criterion directly but we may transform $f(x)$ into a polynomial to which the criterion may be applied.

   Introduce a new indeterminate y where $x = 1 + y$. Then, temporarily using the quotient field $\mathbb{Q}(x) = \mathbb{Q}(y)$ of $\mathbb{Q}[x]$, we have

$$f(1+y) = f(x) = 1 + x + x^2 + \ldots + x^{p-1}$$

$$= \frac{x^p - 1}{x - 1}$$

$$= \frac{(1+y)^p - 1}{y}$$

$$= \frac{\left[1 + \binom{p}{1}y + \binom{p}{2}y^2 + \ldots + \binom{p}{p}y^p\right] - 1}{y}$$

$$= \binom{p}{1}y + \binom{p}{2}y^2 + \ldots + \binom{p}{p}y^p.$$

Now, as we have seen in the proof of Theorem 8, all of the coefficients in $f(1 + y)$ are divisible by p except for $\binom{p}{p} = 1$, also $\binom{p}{1} = p$ and so is not divisible by p^2. Thus $f(1 + y)$ is irreducible and so, evidently, is $f(x)$.

Exercises 5.6

1. Prove that the following polynomials are irreducible in $\mathbb{Q}[x]$.
   $5 + 25x + 10x^2 + 7x^3$, $14 + 21x + 49x^2 + 6x^3$,
   $26 + 39x + 65x^2 + 10x^3$.

2. Prove that $19 + 24x + 9x^2 + x^3$ is irreducible in $\mathbb{Q}[x]$. (Hint: put $x = y - 1$.)

3. Prove that $x^3 + 9$ is irreducible in $\mathbb{Q}[x]$.

Topics in Group Theory

In this final chapter we extend our knowledge of group theory. Among other aspects of finite groups, we investigate permutation groups and obtain two results of cardinal importance in the theory of finite groups. In the first of these, we establish the structure of Abelian groups and, in the second, we establish the existence of the so-called 'Sylow p-subgroups' of a finite group.

6.1 Permutation Groups

We are here concerned with bijective mappings of a non-empty set into itself. Such mappings may be multiplied under the circle-composition of mappings. From our earlier work we easily obtain the following theorem.

Theorem 1

Let X be a non-empty set and let $S(X)$ be the set of bijective mappings of X onto X. Under the circle-composition of mappings $S(X)$ is a group.

Proof

From the Examples in Section 4.1 (Semigroups), we know that $S(X)$ is a monoid. Let $f \in S(X)$. Then the inverse f^{-1} of f is given as follows. f^{-1} is that mapping for which $f^{-1}(a) = b$ if and only if $f(b) = a$ $(a, b \in X)$. It is then immediate that $S(X)$ is a group. □

Our interest in this section concerns bijective mappings on finite sets.

Definition 1

A bijective mapping of a non-empty set X onto itself is called a **permutation**. Any set of permutations of X forming a group is called a **permutation group**. The permutation group of all permutations on X is called the **symmetric group** on X and is frequently denoted by $S(X)$. If X consists of the n **symbols** or **elements** $x_1, x_2, \ldots, x_n$ and more particularly if $X = \{1, 2, \ldots, n\}$ then the symmetric group on X is designated as S_n. A permutation $p \in S_n$ is written as

$$p = \begin{pmatrix} i_1 & i_2 & \cdots & i_n \\ j_1 & j_2 & \cdots & j_n \end{pmatrix}$$

where the notation means that $p(i_1) = j_1$, $p(i_2) = j_2, \ldots, p(i_n) = j_n$. (Usually but not invariably, $i_1, i_2, \ldots, i_n$ are in the natural order $1, 2, \ldots, n$.)

Example 1

The permutation p such that $p(1) = 2, p(2) = 1, p(3) = 4, p(4) = 3$ may be represented by any of

$$\begin{pmatrix} 1 & 2 & 3 & 4 \\ 2 & 1 & 4 & 3 \end{pmatrix}, \begin{pmatrix} 2 & 1 & 4 & 3 \\ 1 & 2 & 3 & 4 \end{pmatrix}, \begin{pmatrix} 4 & 2 & 1 & 3 \\ 3 & 1 & 2 & 4 \end{pmatrix}$$

Definition 2

A permutation $p \in S_n$ is said to **fix** $i \in \{1, 2, \ldots, n\}$ if $p(i) = i$ and to **move** $i \in \{1, 2, \ldots, n\}$ if $p(i) \neq i$. The **identity permutation**, which is the identity of the group S_n, is that permutation fixing all $i \in \{1, 2, \ldots, n\}$.

Lemma 1

S_n is a group of order $n(n-1) \ldots 2.1 = n!$

Proof

Let $p = \begin{pmatrix} 1 & 2 & \cdots & n \\ j_1 & j_2 & \cdots & j_n \end{pmatrix}$.

For j_1 we have n choices of symbols from $\{1, 2, \ldots, n\}$. For j_2 we have $n - 1$ choices of symbols from $\{1, 2, \ldots, n\} \setminus \{j_1\}$. For j_3 we have $n - 2$ choices of symbols from $\{1, 2, \ldots, n\} \setminus \{j_1, j_2\}$. Continuing we have only one choice for j_n. The total number of choices is $n(n - 1)(n - 2) \ldots 1$ giving the result. $\qquad\qquad\square$

Example 2

Let $X = \{1, 2, 3\}$. Then the symmetric group S_3 has order $3! = 6$ and consists of the six permutations,

$$\begin{pmatrix} 1 & 2 & 3 \\ 1 & 2 & 3 \end{pmatrix}, \begin{pmatrix} 1 & 2 & 3 \\ 2 & 3 & 1 \end{pmatrix}, \begin{pmatrix} 1 & 2 & 3 \\ 3 & 1 & 2 \end{pmatrix}, \begin{pmatrix} 1 & 2 & 3 \\ 2 & 1 & 3 \end{pmatrix}, \begin{pmatrix} 1 & 2 & 3 \\ 3 & 2 & 1 \end{pmatrix}, \begin{pmatrix} 1 & 2 & 3 \\ 1 & 3 & 2 \end{pmatrix}.$$

$\begin{pmatrix} 1 & 2 & 3 \\ 1 & 2 & 3 \end{pmatrix}$, is the identity permutation. $\begin{pmatrix} 1 & 2 & 3 \\ 2 & 3 & 1 \end{pmatrix}$ is the permutation p such that $p(1) = 2$, $p(2) = 3$ and $p(3) = 1$.

If $q = \begin{pmatrix} 1 & 2 & 3 \\ 3 & 1 & 2 \end{pmatrix}$, then the product pq is obtained from the composition of the mappings p and q, namely by applying q and then p as follows:

$$1 \xrightarrow{q} 3 \xrightarrow{p} 1$$
$$2 \xrightarrow{q} 1 \xrightarrow{p} 2$$
$$3 \xrightarrow{q} 2 \xrightarrow{p} 3$$

which we also write as

$$\begin{pmatrix} 1 & 2 & 3 \\ 2 & 3 & 1 \end{pmatrix} \begin{pmatrix} 1 & 2 & 3 \\ 3 & 1 & 2 \end{pmatrix} = \begin{pmatrix} 3 & 1 & 2 \\ 1 & 2 & 3 \end{pmatrix} \begin{pmatrix} 1 & 2 & 3 \\ 3 & 1 & 2 \end{pmatrix} = \begin{pmatrix} 1 & 2 & 3 \\ 1 & 2 & 3 \end{pmatrix}.$$

Convention

Let

$$p = \begin{pmatrix} 1 & 2 & \ldots & n \\ i_1 & i_2 & \ldots & i_n \end{pmatrix}, \quad q = \begin{pmatrix} 1 & 2 & \ldots & n \\ j_1 & j_2 & \ldots & j_n \end{pmatrix}.$$

Then

$$p = \begin{pmatrix} 1 & 2 & \ldots & n \\ p(1) & p(2) & \ldots & p(n) \end{pmatrix}, \quad q = \begin{pmatrix} 1 & 2 & \ldots & n \\ q(1) & q(2) & \ldots & q(n) \end{pmatrix}$$

and the product pq is given by

$$
\begin{pmatrix} 1 & 2 & \dots & n \\ p(1) & p(2) & \dots & p(n) \end{pmatrix}
\begin{pmatrix} 1 & 2 & \dots & n \\ q(1) & q(2) & \dots & q(n) \end{pmatrix}
$$

$$
= \begin{pmatrix} q(1) & q(2) & \dots & q(n) \\ (pq)(1) & (pq)(2) & \dots & (pq)(n) \end{pmatrix}
\begin{pmatrix} 1 & 2 & \dots & n \\ q(1) & q(2) & \dots & q(n) \end{pmatrix}
$$

$$
= \begin{pmatrix} 1 & 2 & \dots & n \\ (pq)(1) & (pq)(2) & \dots & (pq)(n) \end{pmatrix}.
$$

(The reader is advised that another convention for the multiplication of permutations is also to be found in some textbooks.)

We give further examples exhibiting our convention for the multiplication of permutations.

Examples 3

1.

$$
\begin{pmatrix} 1 & 2 & 3 & 4 & 5 \\ 1 & 5 & 2 & 4 & 3 \end{pmatrix}
\begin{pmatrix} 1 & 2 & 3 & 4 & 5 \\ 5 & 2 & 1 & 3 & 4 \end{pmatrix}
= \begin{pmatrix} 5 & 2 & 1 & 3 & 4 \\ 3 & 5 & 1 & 2 & 4 \end{pmatrix}
\begin{pmatrix} 1 & 2 & 3 & 4 & 5 \\ 5 & 2 & 1 & 3 & 4 \end{pmatrix}
$$

$$
= \begin{pmatrix} 1 & 2 & 3 & 4 & 5 \\ 3 & 5 & 1 & 2 & 4 \end{pmatrix}.
$$

2.

$$
\begin{pmatrix} 1 & 2 & 3 & 4 & 5 & 6 & 7 \\ 6 & 3 & 7 & 1 & 2 & 4 & 5 \end{pmatrix}
\begin{pmatrix} 1 & 2 & 3 & 4 & 5 & 6 & 7 \\ 5 & 2 & 7 & 3 & 6 & 1 & 4 \end{pmatrix}
$$

$$
= \begin{pmatrix} 5 & 2 & 7 & 3 & 6 & 1 & 4 \\ 2 & 3 & 5 & 7 & 4 & 6 & 1 \end{pmatrix}
\begin{pmatrix} 1 & 2 & 3 & 4 & 5 & 6 & 7 \\ 5 & 2 & 7 & 3 & 6 & 1 & 4 \end{pmatrix}
$$

$$
= \begin{pmatrix} 1 & 2 & 3 & 4 & 5 & 6 & 7 \\ 2 & 3 & 5 & 7 & 4 & 6 & 1 \end{pmatrix}.
$$

Lemma 2

Let $p = \begin{pmatrix} 1 & 2 & \dots & n \\ i_1 & i_2 & \dots & i_n \end{pmatrix}$. Then $p^{-1} = \begin{pmatrix} i_1 & i_2 & \dots & i_n \\ 1 & 2 & \dots & n \end{pmatrix}$.

Proof

$$\begin{pmatrix} i_1 & i_2 & \cdots & i_n \\ 1 & 2 & \cdots & n \end{pmatrix} \begin{pmatrix} 1 & 2 & \cdots & n \\ i_1 & i_2 & \cdots & i_n \end{pmatrix} = \begin{pmatrix} 1 & 2 & \cdots & n \\ 1 & 2 & \cdots & n \end{pmatrix}. \qquad \square$$

Definition 3

A permutation p of $\{1, 2, \ldots, n\}$ is called a **cycle** of length r $(1 < r \le n)$ if for some subset $\{i_1, i_2, \ldots, i_n\}$ from $\{1, 2, \ldots, n\}$ we have $p(i_i) = i_2$, $p(i_2) = i_3, \ldots p(i_{r-1}) = i_r$, $p(i_r) = i_1$ and for $j \in \{1, 2, \ldots, n\} \setminus \{i_1, i_2, \ldots, i_r\}$, $p(j) = j$. To indicate that the cycle p permutes the integers $i_1, i_2, \ldots, i_r$ cyclically and that the remaining integers are fixed by p we write p, in an abbreviated notation, as

$$p = (i_1 i_2 \ldots i_r).$$

A cycle of length 2 is often called a **transposition**. Conventionally the identity permutation is regarded as a cycle of length 1 and is written as (1).

Example 4

$\begin{pmatrix} 1 & 2 & 3 & 4 & 5 & 6 & 7 & 8 \\ 6 & 3 & 5 & 4 & 1 & 2 & 7 & 8 \end{pmatrix}$ is the cycle (1 6 2 3 5), which may also be written as (6 2 3 5 1), (2 3 5 1 6), (3 5 1 6 2) or (5 1 6 2 3).

   We note that in cycle notation we write down only the symbols from $1, 2, \ldots, n$ which are moved. Consequently only the context enables us to determine whether the cycle (1 6 2 3 5) represents $\begin{pmatrix} 1 & 2 & 3 & 4 & 5 & 6 & 7 & 8 \\ 6 & 3 & 5 & 4 & 1 & 2 & 7 & 8 \end{pmatrix}$ or, $\begin{pmatrix} 1 & 2 & 3 & 4 & 5 & 6 \\ 6 & 3 & 5 & 4 & 1 & 2 \end{pmatrix}$ etc. Fortunately ambiguity does not usually arise.

   Permutations expressed in cycle notation are easy to multiply. We illustrate their multiplication in the following example.

Example 5

Suppose we wish to prove that in S_6 we have

$$(1 \ 3 \ 4 \ 5 \ 2)(6 \ 2 \ 3 \ 4)(1 \ 3 \ 5) = (1 \ 5 \ 3 \ 2 \ 4 \ 6).$$

Letting $r = (1 \quad 3 \quad 4 \quad 5 \quad 2)$, $s = (6 \quad 2 \quad 3 \quad 4)$, $t = (1 \quad 3 \quad 5)$ we have, for the product rst,

$$1 \xrightarrow{t} 3 \xrightarrow{s} 4 \xrightarrow{r} 5$$
$$2 \xrightarrow{t} 2 \xrightarrow{s} 3 \xrightarrow{r} 4$$
$$3 \xrightarrow{t} 5 \xrightarrow{s} 5 \xrightarrow{r} 2$$
$$4 \xrightarrow{t} 4 \xrightarrow{s} 6 \xrightarrow{r} 6$$
$$5 \xrightarrow{t} 1 \xrightarrow{s} 1 \xrightarrow{r} 3$$
$$6 \xrightarrow{t} 6 \xrightarrow{s} 2 \xrightarrow{r} 1$$

and so

$$rst = (1 \quad 3 \quad 4 \quad 5 \quad 2)(6 \quad 2 \quad 3 \quad 4)(1 \quad 3 \quad 5) = (1 \quad 5 \quad 3 \quad 2 \quad 4 \quad 6).$$

Example 6

The cycles $(1 \quad 7 \quad 8 \quad 2 \quad 4)$ and $(3 \quad 6 \quad 5)$ permute cyclically the disjoint sets of integers, $\{1, 2, 4, 7, 8\}$ and $\{3, 5, 6\}$. The cycles are easily seen to commute.

Definition 4

Two permutations from S_n are said to be **disjoint** if the two subsets of integers moved by the permutations are disjoint.

Lemma 3

Disjoint permutations commute.

Proof

Let p and q be disjoint permutations from S_n. Then we know that we have three subsets A, B, C, of which C may be empty, such that

$$\{1, 2, \dots, n\} = A \cup B \cup C$$

is a disjoint union and such that p permutes the integers of A among themselves but fixes the integers of $B \cup C$ and q permutes the integers of B among themselves but fixes the integers of $A \cup C$. Let $a \in A$, $b \in B$, $c \in C$. Then

$$(pq)(a) = p(q(a)) = p(a) = q(p(a)) = (qp)(a),$$
$$(pq)(b) = p(q(b)) = q(b) = q(p(b)) = (qp)(b),$$
$$(pq)(c) = p(q(c)) = p(c) = c = q(c) = q(p(c)) = (qp)(c).$$

Thus $pq = qp$. $\qquad\qquad\qquad\qquad\qquad\qquad\qquad\qquad\qquad\qquad\qquad\square$

Example 7

$$\begin{pmatrix} 1 & 2 & 3 & 4 & 5 & 6 & 7 & 8 & 9 \\ 9 & 8 & 3 & 6 & 5 & 7 & 4 & 1 & 2 \end{pmatrix}$$

is a product of the disjoint, and so commuting, cycles $(1\ 9\ 2\ 8)$ and $(4\ 6\ 7)$.

We obtain easily the following result.

Theorem 2

Every permutation is a product of disjoint cycles.

We have broken down any given permutation into disjoint cycles. We may now break down each cycle into transpositions.

Theorem 3

Every permutation is a product of transpositions.

Proof

It is enough to show that every cycle is a product of transpositions.

Trivially $(1) = (12)(12)$. Let now $(i_1, i_2, \ldots, i_r)$ be a cycle of length r, $r > 1$. Then, by direct calculation,

$$(i_1 i_r)(i_1 i_{r-1}) \ldots (i_1 i_3)(i_1 i_2) = (i_1 i_2 \ldots i_r). \qquad \square$$

Example 8

$$\begin{pmatrix} 1 & 2 & 3 & 4 & 5 & 6 & 7 & 8 \\ 7 & 1 & 2 & 3 & 4 & 8 & 5 & 6 \end{pmatrix} = (1\ 7\ 5\ 4\ 3\ 2)(6\ 8)$$

$$= (1\ 2)(1\ 3)(1\ 4)(1\ 5)(1\ 7)(6\ 8)$$

$$= (5\ 7)(5\ 1)(6\ 8)(5\ 2)(1\ 2)(3\ 4)(5\ 3)(1\ 2)$$

We may express a permutation as a product of transpositions in many, indeed too many, ways but whether an even number or an odd number of transpositions is involved in any particular way turns out to be solely determined by the permutation. In order to prove this mysterious fact we make a digression into rings of polynomials of several commuting indeterminates and we define the 'action' of a permutation upon such polynomials.

Definition 5

Let $\mathbb{Z}[x_1, x_2, \ldots, x_n]$ be the polynomial ring of n commuting indeterminates $x_1, x_2, \ldots, x_n$ over $\mathbb{Z}$. Let $p \in S_n$ and let $f(x_1, x_2, \ldots, x_n) \in \mathbb{Z}[x_1, x_2, \ldots, x_n]$. The **action** of p on $f(x_1, x_2, \ldots, x_n)$ is then defined to be the polynomial $(pf)(x_1, x_2, \ldots, x_n)$ given by

$$(pf)(x_1, x_2, \ldots, x_n) = f(x_{p(1)}, x_{p(2)}, \ldots, x_{p(n)}).$$

Example 9

Let $f(x_1, x_2, x_3) = x_1^4 x_2^3 + 5x_1^2 + 4x_2^7 x_3^8 \in \mathbb{Z}[x_1, x_2, x_3]$.

Let $p = \begin{pmatrix} 1 & 2 & 3 \\ 3 & 1 & 2 \end{pmatrix} \in S_3$. Then

$$(pf)(x_1, x_2, x_3) = x_3^4 x_1^3 + 5x_3^2 + 4x_1^7 x_2^8.$$

If, in particular, we act upon $f(x_1, x_2, \ldots, x_n) \in \mathbb{Z}[x_1, x_2, \ldots, x_n]$, as above, by a transposition (ij) $(i \neq j)$, say, then in effect we interchange the indeterminates x_i and x_j in the expression for $f(x_1, x_2, \ldots, x_n)$.

Example 10

We let the transposition $(1 \;\; 2)$ act on various polynomials.

$(1 \;\; 2) (x_1^3 + x_2^2 + x_3) = x_2^3 + x_1^2 + x_3,$

$(1 \;\; 2) (x_1 - x_2) = x_2 - x_1 = -(x_1 - x_2),$

$(1 \;\; 2) (x_3 - x_4) = x_3 - x_4,$

$(1 \;\; 2) (x_1 - x_2)(x_1 - x_3)(x_2 - x_3) = (x_2 - x_1)(x_2 - x_3)(x_1 - x_3)$
$$= -(x_1 - x_2)(x_1 - x_3)(x_2 - x_3).$$

Definition 6

Let $\mathbb{Z}[x_1, x_2, \ldots, x_n]$ be the polynomial ring of n commuting indeterminates over $\mathbb{Z}$. Let $f(x_1, x_2, \ldots, x_n)$ be a polynomial from $\mathbb{Z}[x_1, x_2, \ldots, x_n]$. If for all transpositions $t \in S_n$

$$(tf)(x_1, x_2, \ldots, x_n) = f(x_1, x_2, \ldots, x_n)$$

then $f(x_1, x_2, \ldots, x_n)$ is said to be **symmetric**. If for all transpositions $t \in S_n$

$$(tf)(x_1, x_2, \ldots, x_n) = -f(x_1, x_2, \ldots, x_n)$$

then $f(x_1, x_2, \ldots, x_n)$ is said to be **skew-symmetric**.

A polynomial may be neither symmetric nor skew-symmetric.

Example 11

In $\mathbb{Z}[x_1, x_2, x_3]$, $x_1 + x_2 + x_3$, $x_1^2 + x_2^2 + x_3^2$, $x_1 x_2 + x_1 x_3 + x_2 x_3$ and $x_1 x_2 x_3$ are typical symmetric polynomials. $(x_1 - x_2)(x_1 - x_3)(x_2 - x_3)$ is a skew-symmetric polynomial. $x_1 - x_2 + x_3$ is neither symmetric nor skew-symmetric.

We recall that

$$\prod_{r=1}^{n} a_r$$

denotes the product $a_1 a_2 \ldots a_n$. Let $w(x_1, x_2, \ldots, x_n)$ be the particular polynomial from $\mathbb{Z}[x_1, x_2, \ldots, x_n]$ given by

$$
\begin{aligned}
w(x_1, x_2, \ldots, x_n) &= \prod_{\substack{r,s=1 \\ r<s}}^{n} (x_r - x_s) \\
&= (x_1 - x_2)(x_1 - x_3) \ldots (x_1 - x_n) \\
&\qquad \times (x_2 - x_3) \ldots (x_2 - x_n) \\
&\qquad \times \quad \ldots \\
&\qquad \times (x_{n-1} - x_n)
\end{aligned}
$$

Example 12

$$w(x_1, x_2, x_3, x_4) = (x_1 - x_2)(x_1 - x_3)(x_1 - x_4)(x_2 - x_3)(x_2 - x_4)(x_3 - x_4).$$

Lemma 4

For all transpositions t from S_n

$$(tw)(x_1, x_2, \ldots, x_n) = -w(x_1, x_2, \ldots, x_n).$$

Proof

Let t be the transposition (ij), $i < j$. t acts on the factors of the product

$$\prod_{\substack{r,s=1 \\ r<s}}^{n} (x_r - x_s).$$

A factor of the form $x_r - x_s$ where neither r nor s is equal to either i or j is unaltered by t.

Factors of the form $x_r - x_i$ and $x_r - x_j$ where r is not equal to either i or j are simply interchanged by t. Similarly factors $x_i - x_r$ and $x_j - x_r$, $(r \neq i, j)$ are interchanged by t.

Factors of the form $x_r - x_i$ and $x_j - x_r$ $(r \neq i, j)$ must be considered together. For such a pair of factors we have

$$(ij)[(x_r - x_i)(x_j - x_r)] = (x_r - x_j)(x_i - x_r)$$
$$= (x_j - x_r)(x_r - x_i)$$

and so the product of the pair of factors is unaltered.

The single remaining factor to be considered is $x_i - x_j$ and for this factor

$$(ij)(x_i - x_j) = x_j - x_i = -(x_i - x_j)$$

and so we have a change of sign. Altogether there is, in consequence, effectively one change of sign and hence we obtain the desired result. □

Corollary

Let p be a permutation from S_n. Then

$$(pw)(x_1, x_2, \ldots, x_n) = \pm w(x_1, x_2, \ldots, x_n).$$

Proof

p is a product of transpositions, say $p = t_1 t_2 \ldots t_m$ where $t_1, t_2, \ldots, t_m$ are transpositions. Then

$$(pw)(x_1, x_2, \ldots, x_n) = (t_1 t_2 \ldots t_m w)(x_1, x_2, \ldots, x_n)$$
$$= -(t_1 \ldots t_{m-1} w)(x_1, x_2, \ldots, x_n)$$
$$= \ldots$$
$$= (-1)^m w(x_1, x_2, \ldots, x_n).$$

□

Definition 7

A permutation p from S_n is said to be **even** if

$$(pw)(x_1, x_2, \ldots, x_n) = w(x_1, x_2, \ldots, x_n)$$

and to be **odd** if

$$(pw)(x_1, x_2, \ldots, x_n) = -w(x_1, x_2, \ldots, x_n).$$

By Lemma 4, a transposition is an odd permutation. From the proof of the Corollary above we obtain the next theorem.

Theorem 4

Every permutation is either even or odd. An even permutation is only expressible as a product of an even number of transpositions. An odd permutation is only expressible as a product of an odd number of permutations.

We may go further as follows.

Theorem 5

Let A_n be the set of even permutations from S_n. Then A_n is a normal subgroup of S_n of index 2, $|A_n| = \frac{1}{2}n!$

Proof

A_n is obviously closed under multiplication and if $p \in A_n$ then p is a product of an even number of transpositions and so also is p^{-1}. Thus A_n is a group.

We claim that $S_n = A_n \cup A_n(1\ 2)$. Let $p \in S_n$. If p is even then $p \in A_n$. If p is odd then $p(1\ 2)$ is even and so $p(1\ 2) \in A_n$ from which $p \in A_n(1\ 2)$. Therefore $S_n \subseteq A_n \cup A_n(1\ 2)$ giving the result. □

Definition 8

The group of the even permutations from S_n is called the **alternating group** on $\{1, 2, \ldots, n\}$ and is denoted by A_n.

Example 13

$$A_3 = \{(1), (1\ \ 2\ \ 3), (1\ \ 3\ \ 2)\}.$$

A situation which commonly arises, and which we have encountered, is that in which a group G 'acts' as a permutation group on a non-empty set X. We sometimes say that G 'induces' a permutation group on X. We require to make this terminology more precise.

Definition 9

Let G be a group and let X be a non-empty set. Let $S(X)$ be the group of permutations of X and let $f : G \to S(X)$ be a homomorphism. Then for each $a \in G$, $f(a)$ is a permutation of X and G is then said to **act** as a permutation group on X or to **induce** a permutation group on X. As a short notation for the action of G on X we write

$$a.x = f(a)x \quad (x \in X, a \in G).$$

We note that, in this notation,

$$(ab).x = f(ab)x = (f(a)f(b))x = f(a)(f(b)x) = a.(b.x) \quad (x \in X, a, b \in G).$$

Furthermore

$$\text{Ker } f = \{a \in G \,|\, a.x = x \text{ for all } x \in X\}$$

and so $G/\text{Ker } f$ is isomorphic to a subgroup of $S(X)$.

Definition 10

Let G act as a permutation group on a non-empty set X. Let $x \in X$. Then the **orbit** containing x is defined to be $G(x) = \{a.x \,|\, a \in G\}$. The **stabilizer** of x is defined to be

$$\text{Stab}_G(x) = \{a \in G \,|\, a.x = x\}.$$

Theorem 6

Let G act as a permutation group on a non-empty set X. A relation $\sim$ is defined on X by $x \sim y$ $(x, y \in X)$ if and only if $a.x = y$ for some $a \in G$.

1. Then $\sim$ is an equivalence relation and each equivalence class is an orbit.
2. If X is finite the number of distinct elements in the orbit of $x \in X$ is given by $|G : \text{Stab}_G(x)|$.

Proof

1. Certainly for all $x \in G$, $e.x = x$ and so $x \sim x$.

   If $x \sim y$ $(x, y \in X)$ then there exists $a \in G$ such that $a.x = y$. Hence it follows that $x = e.x = (a^{-1}a).x = a^{-1}.(a.x) = a^{-1}.y$ and so $y \sim x$.

   If $x \sim y$ and $y \sim z$ $(x, y, z \in G)$ then there exist $a, b \in G$ such that $a.x = y$ and $b.y = z$, from which $(ba).x = b.(a.x) = b.y = z$ and so $x \sim z$.

2. $G(x) = \{a.x \,|\, a \in G\}$ is the orbit of x. Let

$$G = a_1\text{Stab}_G(x) \cup a_2\text{Stab}_G(x) \cup \ldots \cup a_m\text{Stab}_G(x)$$

be a coset decomposition of $\text{Stab}_G(x)$ in G. Then we claim that the elements $a_1.x, a_2.x, \ldots, a_m.x$ are distinct and constitute the entire orbit containing x. The proof is very similar to that in Chapter 4, Theorem 19, in which we proved that the number of elements in a conjugacy class of a group is equal to the index of the centralizer of an element in the class. The group acts as a permutation group on the conjugacy class. The reader should establish for himself or herself the details of the present proof.   □

We may now prove a result first established by Cayley in 1878.

Theorem 7 Cayley's Theorem

A finite group of order n is isomorphic to a subgroup of the symmetric group S_n.

Proof

Let G have distinct elements $x_1, x_2, \ldots, x_n$. Then G acts as a permutation group on G where G is considered simply as a non-empty set. The action is given by

$$a.x_i = ax_i \quad (a \in G, i = 1, 2, \ldots, n)$$

since we have

$$ab.x_i = (ab)x_i = a(bx_i) = a.(b.x_i) \quad (a, b \in G, i = 1, 2, \ldots, n).$$

Furthermore $a \in G$ acts as the identity permutation on X if and only if $ax_i = a.x_i = x_i$ $(i = 1, 2, \ldots, n)$ which is so if and only if a is the identity of G. Thus G is isomorphic to a subgroup of the symmetric group on $\{x_1, x_2, \ldots, x_n\}$. [In fact the isomorphism is given explicitly by

$$a \to \begin{pmatrix} x_1 & x_2 & \cdots & x_n \\ ax_1 & ax_2 & \cdots & ax_n \end{pmatrix}.]$$ □

Exercises 6.1

1. Evaluate the following products:

$$\begin{pmatrix} 1 & 2 & 3 & 4 & 5 & 6 \\ 3 & 1 & 4 & 2 & 6 & 5 \end{pmatrix}\begin{pmatrix} 1 & 2 & 3 & 4 & 5 & 6 \\ 5 & 4 & 6 & 3 & 2 & 1 \end{pmatrix},$$

$$\begin{pmatrix} 1 & 2 & 3 & 4 & 5 & 6 \\ 2 & 1 & 4 & 3 & 6 & 5 \end{pmatrix}\begin{pmatrix} 1 & 2 & 3 & 4 & 5 & 6 \\ 1 & 3 & 4 & 5 & 6 & 2 \end{pmatrix}.$$

2. Prove that

   $(1 \quad 2 \quad 9 \quad 3 \quad 4)(5 \quad 6 \quad 2 \quad 8 \quad 1)(7 \quad 8 \quad 9 \quad 1)$

   $= (1 \quad 7 \quad 2 \quad 8 \quad 3 \quad 4)(5 \quad 6 \quad 9).$

3. Write the following cycles as products of transpositions:

   $(1 \quad 3 \quad 5 \quad 7 \quad 6 \quad 2), \quad (1 \quad 2 \quad 4 \quad 3), \quad (6 \quad 7 \quad 1 \quad 2 \quad 5 \quad 8 \quad 4).$

   Which cycles are even and which are odd?

4. Write down the elements of the A_4.

5. Is the polynomial $(x_1 - x_2)(x_1 - x_3)(x_2 - x_3)(x_1 + x_2 + x_3)$ symmetric, skew-symmetric or neither?

6. Let G be a group and let $f : G \to S(X)$ be a homomorphism of G into the group $S(X)$ of permutations on a finite set X. Prove that

$$\operatorname{Ker} f = \bigcap_{x \in X} \operatorname{Stab}_G(x).$$

7. Write down the full proof of Theorem 6, part 2.

8. Let $X = \{H_1, H_2, \ldots, H_n\}$ be a complete set of conjugates of a subgroup H of a group G. Prove that an action of G on X is defined by

$$a.H_i = aH_i a^{-1} \quad (a \in G, i = 1, 2, \ldots, n).$$

Prove that $n = |G : N_G(H)|$.

9. Find an isomorphism between the S_3 and the non-Abelian of order 6, the Cayley table of which is in the Examples 11 no. 3 of Section 4.2.

6.2 Generators and Relations

When we discussed various finite groups in Chapter 4 we often wrote down their Cayley tables. It was evident that these tables were only useable for groups of order perhaps at most 20 and so their use had obvious limitations for describing finite groups in general. We here develop a more efficient means of describing groups, whether finite or infinite.

Example 14

Let us consider the group G of order 6 given in Examples 11 no. 3 of Section 4.2, in which $G = \{e, a, b, c, d, f\}$ is the group of mappings of $\mathbb{R} \setminus \{0, 1\}$ into $\mathbb{R} \setminus \{0, 1\}$ for which the Cayley table has been given.

We have $a^3 = e$, $c^2 = e$. Suppose we try to find the least subgroup H of G containing a and c. Certainly H contains e, a, a^2, c, $d = ac$ and $f = a^2 c$ and so $H = G$. Thus we say that a and c 'generate' G. Could we determine the Cayley table from simply 'knowing' a and c? The answer is 'not quite', because a and c are not unrelated since we have $ac = ca^2 (= d)$ and $a^2 c = ca (= f)$. But these relations may be rewritten as $c^{-1}ac = a^2$, $c^{-1}a^2 c = a$ and the second is a consequence of the first since $c^{-1}a^2 c = (c^{-1}ac)^2 = (a^2)^2 = a^4 = a$. But now knowing the fact that in G we have $a^3 = e$, $c^2 = e$, $c^{-1}ac = a^2$ we may work out the products of elements such as df by using only this fact and by using the legitimate operations of a group; thus we may determine df as follows. We have $d = ac$ and $f = a^2c$ and so $df = (ac)(a^2c) = ac^{-1}a^2c = aa = b$. We can obtain all the entries of the Cayley table in this way. Thus we may derive the Cayley table from knowing that we have a group generated by two elements a and c with the relations $a^3 = e$, $c^2 = e$, $c^{-1}ac = a^2$. We write

$$G = \langle a, c \,|\, a^3 = e,\, c^2 = e,\, c^{-1}ac = a^2 \rangle.$$

Example 15

Suppose we consider a group G with generators a and c as in the previous example but with a slight modification of the relations, namely we suppose $a^3 = e$, $c^2 = e$, $c^{-1}ac = a$.

By virtue of the fact that a and c commute, we may collect together in any product of a's and c's the powers of a and c. Using the relations above we see that $G = \{e, a, a^2, c, ac, a^2c\}$ and is an Abelian group of order 6. We may write down a Cayley table as shown below:

	e	a	a^2	c	ac	a^2c
e	e	a	a^2	c	ac	a^2c
a	a	a^2	e	ac	a^2c	c
a^2	a^2	e	a	a^2c	c	ac
c	c	ac	a^2c	e	a	a^2
ac	ac	a^2c	c	a	a^2	e
a^2c	a^2c	c	ac	a^2	e	a

We deduce more. If we let $b = ac$ then $b^2 = a^2$, $b^3 = c$, $b^4 = a$, $b^5 = a^2c$, $b^6 = e$. We see that G is actually a cyclic group of order 6 which may be represented as either

$$G = \langle a, c \,|\, a^3 = e,\, c^2 = e,\, c^{-1}ac = a \rangle$$

or as

$$G = \langle b \,|\, b^6 = e \rangle.$$

Thus, for a given group, generators and relations are not uniquely determined.

Remark

In both of the examples above there were two generators a and c for which $a^3 = e$, $c^2 = e$ and one further relation between a and c. We take this opportunity to remark that if we merely have two elements a and c for which $a^3 = e$,

$c^2 = e$ and no relation is presumed to exist between a and c then it may be shown, but regrettably not easily, that the group so described is isomorphic to the group of linear fractional transformations of the form

$$z \rightarrow \frac{\alpha z + \beta}{\gamma z + \delta} \quad (\alpha, \beta, \gamma, \delta \in \mathbb{Z},\ \alpha\delta - \beta\gamma = 1)$$

where now a and c correspond respectively to the transformations

$$z \rightarrow -\frac{1}{z+1}, \quad z \rightarrow \frac{1}{z}.$$

In the two examples above we began with a group, the elements of which could be written down or easily obtained, and we have shown how by considering a few (actually, two) of these elements and some relations between them we may, in some sense, 'recover' the group. What is much more usual is to begin with some elements supposed to be the elements of a group and with some relations between those elements and then to investigate the group which these generators and relations describe. We do not assert either that the elements are necessarily distinct or that the relations are completely independent of one another. In these circumstances the reader may wonder whether the resulting algebraic entity is, genuinely, a group. The answer is fortunately in the affirmative since the group of one element $\{e\}$ will always satisfy any system of generators and relations albeit trivially. The real interest lies in determining the 'largest' group, up to isomorphism, satisfying the system.

Every group has at least one system of generators and relations. We need only write down the elements $a, b, c, \ldots$ and the relations formed of products such as $ab = c$ to get a system of generators and relations, but such a system is no more efficient than a Cayley table. We aim to be more economical with any generators and relations for a group.

Notation

The group G with **generators** $a_1, a_2, \ldots$ and **relations** $r_1 = s_1, r_2 = s_2, \ldots$ where $r_1, r_2, \ldots;\ s_1, s_2, \ldots$ involve only $a_1, a_2, \ldots$ is denoted by

$$G = \langle a_1, a_2, \ldots | r_1 = s_1, r_2 = s_2, \ldots \rangle.$$

(We shall restrict our attention in this text to a finite number of generators and relations.)

Example 16

Let $G = \langle a, b, c \,|\, a^2 = b^2 = c^3 = e, ab = ba = c \rangle$. What is the group G? We have $e = a^2 b^2 = aabb = abab = (ab)^2 = c^2$. But $c^3 = e$ and so $c = e$. Then $b = a^{-1} = a$. Thus, in truth, G is a group of two elements, namely e and a, and $a^2 = e$. Consequently $G = \langle a \,|\, a^2 = e \rangle$.

We now consider some standard examples.

Examples 17

1. Let $G = \langle a \,|\, a^n = e \rangle$. Then G is the cyclic group of order n generated by a.

2. Let $G = \langle a \,|\, . \rangle$. Then G is the group generated by a with an empty set of relations. Thus $e, a, a^{-1}, a^2, a^{-2}, \ldots$ are distinct and G is the infinite cyclic group generated by a.

3. Let $G = \langle a, b \,|\, a^2 = b^2 = e, ab = ba \rangle$. Then G is isomorphic to the Klein four-group.

4. Let $G = \langle b, f \,|\, b^4 = e, f^2 = b^2, f^{-1}bf = b^3 \rangle$. Then we see that $\langle b \rangle$, the subgroup generated by b is normal in G and has order 4. $\langle b \rangle f \neq \langle b \rangle$ but since $f^2 = b^2$ and f^2 is in the centre of G, no coset distinct from either $\langle b \rangle$ or $\langle b \rangle f$ is possible. Thus G has order 8. In fact G is the quaternion group of Exercises 4.5, no. 5 with the same b and f and with $a = b^2$, $c = b^3$, $h = bf$, $d = b^2 f$, $g = b^3 f$.

5. Let $G = \langle a, b \,|\, a^2 = b^2 = e \rangle$. Then G is generated by a and b, both of order 2, but no relation connects a and b. The structure of G is not easy to discern but becomes clearer if we make a small change to the generators. Let $c = ab$. Then we have $c^{-1} = (ab)^{-1} = b^{-1}a^{-1} = ba$ and then $b^{-1}cb = babb = ba = c^{-1}$. Hence it follows that $G = \langle b, c \,|\, b^2 = e, b^{-1}cb = c^{-1} \rangle$ and so G has a normal subgroup $\langle c \rangle$ which is an infinite cyclic subgroup of index 2. The group is known as the **infinite dihedral group**.

Groups defined by generators and relations appear in many contexts. One context in particular is that of a group of transformations of some regular geometrical object, the transformations preserving the size, shape and location of the object. Under such a group the vertices of the object are permuted amongst themselves and the group is directly a subgroup of the symmetric group on the set of vertices.

Consider, for example, the group of transformations of a square with centre 0 and vertices labelled 1, 2, 3, 4. We may rotate the square about its centre 0 through angles which are multiples of $\dfrac{2\pi}{4} = \dfrac{\pi}{2}$. The permutation $\rho = (1\ 2\ 3\ 4)$ represents a (clockwise) rotation through an angle of $\dfrac{\pi}{2}$.

The remaining rotations about 0 are $\rho^2, \rho^3, \rho^4 = e$ where e is the identity transformation ($e = (1)$). The square may also be reflected about an axis of symmetry, say about a line through 0, parallel to a side, as shown below.

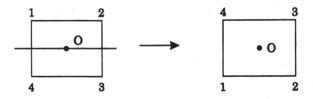

The reflection τ is given by $\tau = (1\ 4)(2\ 3)$. We now claim that the group G of all such transformations is generated by p and τ. Certainly $\rho^4 = e$, $\tau^2 = e$ and

$$\tau^{-1}\rho\tau = (1\ 4)(2\ 3)(1\ 2\ 3\ 4)(1\ 4)(2\ 3) = (1\ 4\ 3\ 2) = \rho^{-1},$$

and so

$$G = \langle \rho, \tau \,|\, \rho^4 = \tau^2 = e,\ \tau^{-1}\rho\tau = \rho^{-1}\rangle$$

but we have to establish that every transformation is indeed in G. We confine our considerations to just one further transformation, leaving the reader to ponder the remaining possibilities. We examine the reflection $(1\ 3)$ about the diagonal joining 2 and 4.

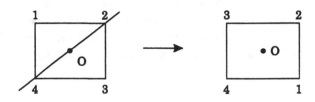

Then $(1\ 3) = (1\ 4\ 3\ 2)(1\ 4)(2\ 3) = \rho^{-1}\tau \in G$. We have now proved that the group of transformations preserving a square is a group which previously we called the dihedral group D_4 of order 8.

We may extend these arguments to consider the group of transformations of a regular polygon. If the polygon has n vertices, labelled $1, 2, \ldots, n$, then it may be rotated through angles which are multiples of $\dfrac{2\pi}{n}$, thus yielding the transformation $\rho = (1\ 2 \ldots n)$ and powers of ρ. The polygon may also be reflected about an axis of symmetry through its centre.

If n is even, $n = 2N$ (say), the axis of symmetry may be chosen (as in the case of the square) to be the perpendicular bisector of the side joining 1 and $2N$. We then have a reflection $\tau = (1\ 2N)(2\ 2N-1)\ldots(N\ N+1)$. As $\rho = (1\ 2 \ldots 2N)$ we have

$$\rho^{2N} = \tau^2 = e, \quad \tau^{-1}\rho\tau = \rho^{-1}.$$

If n is odd, $n = 2N+1$ (say), the axis of symmetry must pass through one vertex $2N+1$ (say) and we have a reflection $\tau = (1\ 2N)(2\ 2N-1)\ldots (N\ N+1)$. Since $\rho = (1\ 2 \ldots 2N+1)$ we have $\rho^{2N+1} = \tau^2 = e$, $\tau^{-1}\rho\tau = \rho^{-1}$.

In both cases we may show that the group generated by ρ and τ contains all transformations of the polygon. In summary, we have shown that the group of transformations of a regular **n-gon** is

$$\langle \rho, \tau \,|\, \rho^n = \tau^2 = e, \tau^{-1}\rho\tau = \rho^{-1} \rangle$$

which is called the **dihedral group** D_n of order $2n$.

Exercises 6.2

1. Let $G = \langle a, b \,|\, a^3 = b^2 = e, ba = e \rangle$. What is the order of G?

2. Let $G = \langle a, b \,|\, a^4 = b^2 = e, aba^2 = b \rangle$. Prove that G has order 2.

3. Let $G = \langle a, b \,|\, a^8 = e, a^4 = b^2, b^{-1}ab = a^{-1} \rangle$. Prove that G has order 16 and that G has only one subgroup of order 2.

4. Prove that $G = \langle a, b \,|\, a^3 = b^2 = e, b^{-1}ab = a^{-1} \rangle$ is isomorphic to the S_3 and that G is also the group of transformations preserving size etc. of an equilateral triangle. (See Exercises 6.1, no. 9.)

5. What is the group of transformations preserving size etc. of a rectangle which is not a square?

6. Prove that all transformations preserving size etc. of a regular pentagon may be identified with elements of the dihedral group D_5.

6.3 Direct Products and Sums

Let G be the cyclic group of order 6 generated by a,

$$G = \langle a | a^6 = e \rangle.$$

G has two proper subgroups, namely $P = \{e, a^3\}$ and $Q = \{e, a, a^2\}$ of orders 2 and 3 respectively. These subgroups are normal in G and we obviously have $P \cap Q = \{e\}$ and $P \cap Q = G$. We describe this situation more generally in the following definition.

Definition 11

Let G be a group. Let H and K be normal subgroups of G such that the intersection $H \cap K = \{e\}$ and $G = HK$. Then G is said to be the (internal) direct product of H and K.

Let us now consider two groups A and B where $A = \langle a | a^2 = e_A \rangle$ and $B = \langle b | b^3 = e_B \rangle$, e_A and e_B being the identities of A and B, and A and B are cyclic groups of orders 2 and 3 respectively. Let G be the Cartesian product $A \times B$ of A and B, then

$$G = \{(e_A, e_B), (e_A, b), (e_A, b^2), (a, e_B), (a, b), (a, b^2)\}.$$

Let the multiplication in G be component-wise so that, for example,

$$(e_A, b)(a, b) = (e_A a, bb) = (a, b^2).$$

Then under this multiplication it may be seen that $A \times B$ is a group. Again, more generally, we describe this situation in a definition.

Definition 12

Let G be the Cartesian product of the groups A and B. A multiplication is defined on $G = A \times B$ by

$$(a, b)(a', b') = (aa', bb').$$

Then G is a group called the (external) direct product of A and B.

In the group $G = A \times B$ we notice that

$$\bar{A} = \{(a, e_B) | a \in A\} \text{ and } \bar{B} = \{(e_A, b) | b \in B\}$$

are normal subgroups of G and that the mappings

$$a \to (a, e_B) \quad (a \in A) \text{ and } b \to (e_A, b) \quad (b \in B)$$

yield isomorphisms of A with $\bar{A}$ and B with $\bar{B}$ respectively. Moreover G is the (internal) direct product of $\bar{A}$ and $\bar{B}$.

Both of the definitions above extend naturally to direct products involving a finite number of subgroups or groups.

Definition 13

Let G be a group. Let $H_1, H_2, \ldots, H_n$ be normal subgroups of G such that

$$H_i \cap H_1 H_2 \ldots H_{i-1} H_{i+1} \ldots H_n = \{e\} \quad (i = 1, 2, \ldots, n) \text{ and } G = H_1 H_2 \ldots H_n.$$

Then G is said to be the (**internal**) **direct product** of $H_1, H_2, \ldots, H_n$.

In this definition, since $H_1, H_2, \ldots, H_n$ are normal subgroups of G, it follows that, for each i, $1 \leq i \leq n$, $H_1 H_2 \ldots H_{i-1} H_{i+1} \ldots H_n$ is a normal subgroup of G.

Definition 14

Let G be the Cartesian product of the groups $A_1, A_2, \ldots, A_n$. A multiplication is defined on $G = A_1 \times A_2 \times \ldots \times A_n$ by

$$(a_1, a_2, \ldots, a_n)(a_1', a_2', \ldots, a_n') = (a_1 a_1', a_2 a_2', \ldots, a_n a_n')$$

$$(a_i, a_i' \in A_i, i = 1, 2, \ldots, n)$$

Then G is a group called the (**external**) **direct product** of $A_1, A_2, \ldots, A_n$.

Similarly, as above,

$$\bar{A}_i = \{(e_1, e_2, \ldots, e_{i-1}, a, e_{i+1}, \ldots e_n) \mid a \in A_i\},$$

where A_j has the identity e_j $(j = 1, 2, \ldots, n)$, is isomorphic to A_i $(i = 1, 2, \ldots, n)$ and G is the (internal) direct product of $\bar{A}_1, \bar{A}_2, \ldots, \bar{A}_n$. On account of these obvious isomorphisms we sometimes do not always draw a precise distinction between internal and external direct products.

Definition 15

Let p be a prime. A group the order of which is a power of p is called a **p-group**.

Definition 16

Let p be a prime. Let $H_1, H_2, \ldots, H_n$ be cyclic groups of order p. Any group isomorphic to the direct product $H_1 \times H_2 \times \ldots \times H_n$ is said to be an **elementary p-group**.

Example 18

The Klein four-group is an elementary Abelian 2-group.

We have given the definitions above for groups having a binary operation of multiplication. If we are concerned with Abelian groups then, as we know, an additive notation is commonly employed. The changes of notation and nomenclature are fairly obvious. We may still form the Cartesian product of Abelian groups to obtain an additive group in which addition is performed componentwise, yielding an **(external) direct sum**. We find it to be convenient now to rephrase the definition above for **(internal) direct sums** of Abelian groups.

Definition 17 (following Definition 13)

Let A be an Abelian group. Let $A_1, A_2, \ldots, A_n$ be subgroups of A such that

$$A_i \cap (A_1 + A_2 + \ldots + A_{i-1} + A_{i+1} + \ldots + A_n) = \{0\} \quad (i = 1, 2, \ldots, n)$$

and $A = A_1 + A_2 + \ldots + A_n$. Then A is the **(internal) direct sum** of $A_1, A_2, \ldots, A_n$.

Examples 19

1. Let A, B and C be groups of orders 9, 14 and 18 respectively. Then the group G, given by $G = A \times B \times C$, is a group of order $9.14.18 = 2268$.

2. Let A, B, C be cyclic groups of orders $7, 8$ and 17 respectively. Then the group G, given by $G = A \times B \times C$ is a group of order 952. Moreover if

$$A = \langle a | a^7 = e_A \rangle, \ B = \langle b | b^8 = e_B \rangle \text{ and } C = \langle c | c^{17} = e_C \rangle$$

then G is cyclic with generator (a, b, c) and identity element (e_A, e_B, e_C).

Exercises 6.3

1. Let A, B and C be normal subgroups of a group G. Prove that ABC is a normal subgroup of G.

2. Let the group G be the direct product of the groups H and K. Prove that G is Abelian if and only if H and K are Abelian.

3. Let H and K be normal subgroups of a group G such that $G = HK$. Prove that $G/(H \cap K)$ is the direct product of $H/(H \cap K)$ and $K/(H \cap K)$.

4. Let A, B and C be subgroups of an Abelian group G such that

$$G = A + B + C, \ A \cap B = \{0\}, \ A \cap C = \{0\} \text{ and } B \cap C = \{0\}.$$

Does it follow that G is the direct sum of A, B and C?

6.4 Abelian Groups

The structure of an Abelian group may be very complicated and, without some conditions on the group, may be difficult to determine. If the group has a finite number of generators then a structure theorem is possible. Here we shall further restrict the considerations to finite Abelian groups for which we obtain a structure theorem.

We have shown that a cyclic group has subgroups of all possible orders and that the order of an element in a group is the order of the cyclic subgroup generated by that element. If the order of every element of a group is a power of a prime p then it is not immediate, at present, that the group is a p-group since the order might be divisible by some prime other than p and yet have no elements of that order. One of our aims is to eliminate this hypothetical situation.

We begin with a lemma which resolves the situation for an Abelian group. As is customary we shall employ additive notation when discussing Abelian groups.

Lemma 5

Let A be a finite Abelian group. Let p be a prime such that every element of A has an order which is a power of p. Then A is a p-group.

Proof

If A is cyclic then, by the remarks above, A is a p-group since if $|A|$ is divisible by a prime q, $q \neq p$, we would have an element of order q, which is false.

We may now argue by induction and assume the result is true for all groups of orders strictly less than $|A|$. Suppose A is not cyclic and let $a \in A$, $a \neq 0$. Then the subgroup $\langle a \rangle$ generated by a has an order which is a power of p. Further, $A/\langle a \rangle$ has the property that every element has an order which is a power of p since if $x \in A$ and x has order p^α then

$$p^\alpha(x + \langle a \rangle) = p^\alpha x + \langle a \rangle = 0 + \langle a \rangle = \langle a \rangle$$

and so $x + \langle a \rangle$ has order dividing p^α. By the induction assumption $A/\langle a \rangle$ is a p-group and so, as $|A| = |\langle a \rangle| = |\langle a \rangle||A/\langle a \rangle|$, A is a p-group. This completes the proof. $\square$

Theorem 8

Let A be a group of order where $p_1^{\alpha_1} p_2^{\alpha_2} \ldots p_n^{\alpha_n}$ where $p_1, p_2, \ldots, p_n$ are distinct primes and $\alpha_i > 0$ $(i = 1, 2, \ldots, n)$. Let $A_i = \{x \in A \mid p_i^{\alpha_i} x = 0\}$

$(i = 1, 2, \ldots, n)$. Then A_i is a subgroup of order $p_i^{\alpha_i}$ $(i = 1, 2, \ldots, n)$ and A is the direct sum of $A_1, A_2, \ldots, A_n$.

Proof

By Lemma 5, A_i is a p_i-subgroup of A $(i = 1, 2, \ldots, n)$. We want to show that

$$A = A_1 + A_2 + \ldots + A_n,$$

$$A_i \cap (A_1 + A_2 + \ldots + A_{i-1} + A_{i+1} + \ldots + A_n) = \{0\} \quad (i = 1, 2, \ldots, n)$$

and finally that $|A_i| = p_i^{\alpha_i}$ $(i = 1, 2, \ldots, n)$.

Let q_i be given by $|A| = p_i^{\alpha_i} q_i$ $(i = 1, 2, \ldots, n)$. Since $p_1, p_2, \ldots, p_n$ are distinct primes, $q_1, q_2, \ldots, q_n$ have 1 as their greatest common divisor. Hence there exist $t_1, t_2, \ldots, t_n \in \mathbb{Z}$ such that

$$t_1 q_1 + t_2 q_2 + \ldots + t_n q_n = 1.$$

Then, if $x \in A$,

$$x = t_1 q_1 x + t_2 q_2 x + \ldots + t_n q_n x \in A_1 + A_2 + \ldots + A_n$$

since $p_i^{\alpha_i} q_i x = |A| x = 0$ implies that $q_i x \in A_i$ $(i = 1, 2, \ldots, n)$.

Let now

$$y \in A_j \cap (A_1 + A_2 + \ldots + A_{j-1} + A_{j+1} + \ldots + A_n)$$

for some $j \in \{1, 2, \ldots, n\}$. Since $y \in A_j$ we have $p_j^{\alpha_j} y_j = 0$. On the other hand, since $y \in A_1 + A_2 + \ldots + A_{j-1} + A_{j+1} + \ldots + A_n$ we have

$$y = y_1 + y_2 + \ldots + y_{j-1} + y_{j+1} + \ldots + y_n$$

where $y_k \in A_k$ $(k = 1, 2, \ldots, j-1, j+1, \ldots n)$. Then

$$q_j y = q_j y_1 + q_j y_2 + \ldots + q_j y_{j-1} + q_j y_{j+1} + \ldots + q_j y_n = 0 + 0 + \ldots + 0 = 0.$$

Thus the order of y divides the coprime integers p_j^{α} and q_j and so $y = 0$. Thus

$$A_j \cap (A_1 + A_2 + \ldots + A_{j-1} + A_{j+1} + \ldots + A_n) = \{0\} \quad (j = 1, 2, \ldots, n).$$

Hence A is the direct sum of $A_1, A_2, \ldots, A_n$, and so

$$p_1^{\alpha_1} p_2^{\alpha_2} \ldots p_n^{\alpha_n} = |A| = |A_1| |A_2| \ldots |A_n|.$$

Since $|A_i|$ is a power of p_i $(i = 1, 2, \ldots, n)$ we deduce that $|A_i| = p_i^{\alpha_i}$ $(i = 1, 2, \ldots, n)$ and our proof is now complete. $\qquad \square$

By virtue of this last theorem we need only consider the structure of a finite Abelian p-group. The following rather obvious lemma is useful.

Lemma 6

Let G be a group. Let H_1, H_2 and K be subgroups of G such that K is a normal subgroup of G and $K \subseteq H_1 \cap H_2$. If $H_1/K \cap H_2/K$ is the trivial subgroup of G/K then $K = H_1 \cap H_2$.

Proof

Let $x \in H_1 \cap H_2$. Then $Kx \in H_1/K \cap H_2/K$ and so, by assumption, $Kx = K$. Thus $x \in K$ giving the result. $\square$

Lemma 7

Let A be a finite Abelian p-group. Let a be an element of A of greatest possible order. Then there exists a subgroup H of A such that A is the direct sum of H and $\langle a \rangle$.

Proof

If A is cyclic then $A = \langle a \rangle$ and the result is true with $H = \{0\}$. Suppose A is not cyclic. We shall argue by induction on the order of A, assuming that the result is true for groups of orders strictly less than $|A|$.

Then a is of greatest possible order p^m, say, in A and $A \neq \langle a \rangle$. We claim that there exists in $G \setminus \langle a \rangle$ an element of order p. We choose $b \in G \setminus \langle a \rangle$ to have least possible order amongst the elements of $G \setminus \langle a \rangle$. If $pb = 0$ then our claim is proved. If $pb \neq 0$ then b has order p^r where $p < p^r \leq p^m$. But then pb has order p^{r-1} and so, by the choice of b, $pb \in \langle a \rangle$. Thus $pb = na$ for some $n \in \mathbb{Z}$ and hence

$$0 = p^r b = p^{r-1}(pb) = p^{r-1}(na) = (p^{r-1}n)a.$$

Since a has order p^m it must be that p^m divides $p^{r-1}n$ and so p divides n. Thus $n = pq$ for some $q \in \mathbb{Z}$. Let $c = b - qa$. Since $b \notin \langle a \rangle$, we necessarily have $c \notin \langle a \rangle$. But $pc = pb - pqa = pb - na = 0$ and so $c \in A \setminus \langle a \rangle$ and c has order p. This proves our claim.

We may now suppose that b has order p. Then $\langle b \rangle \cap \langle a \rangle = \{0\}$ since $\langle b \rangle \cap \langle a \rangle \neq 0$ implies $\langle b \rangle \subseteq \langle a \rangle$ which is false. Then $a + \langle b \rangle$ has order p^m in $A/\langle b \rangle$ and so $a + \langle b \rangle$ is an element of greatest possible order in $A/\langle b \rangle$. By the induction assumption there exists a subgroup such that $A/\langle b \rangle$ is the direct sum of this subgroup and $(\langle a \rangle + \langle b \rangle)/\langle b \rangle$. We may write the subgroup in the form $H/\langle b \rangle$ for some subgroup H of A.

Then, as $\langle b \rangle \subseteq H$, $A = H + (\langle a \rangle + \langle b \rangle) = H + \langle a \rangle$. In addition we have $H \cap (\langle a \rangle + \langle b \rangle) = \langle b \rangle$.

To complete the proof we have to show that $H \cap \langle a \rangle = \{0\}$. Let, therefore, $x \in H \cap \langle a \rangle$. Then $x \in H \cap (\langle a \rangle + \langle b \rangle) = \langle b \rangle$. Thus $x \in \langle a \rangle \cap \langle b \rangle = \{0\}$ and we have established the lemma. $\qquad \square$

Corollary

A finite Abelian p-group is a direct sum of cyclic p-subgroups.

Proof

Let A be a finite Abelian p-group. By Lemma 7 there exists a cyclic subgroup A_1 and a subgroup H such that

$$A = A_1 + H, \quad A_1 \cap H = \{0\}.$$

Now H has strictly lower order than A and so a simple induction ensures that H is a direct sum of cyclic p-groups $A_2, A_3, \ldots, A_n$, say.

Thus

$$A = A_1 + (A_2 + \ldots + A_3) = A_1 + A_2 + \ldots + A_3.$$

and

$$A_1 \cap (A_2 + A_3 + \ldots + A_n) = \{0\}.$$

and so, finally, A is the direct sum of $A_1, A_2, \ldots, A_n$. $\qquad \square$

We are now able to prove our main theorem on the structure of finite Abelian groups.

Theorem 9 Fundamental Theorem of Finite Abelian Groups

A finite Abelian p-group A is direct sum of cyclic p-subgroups. If A is a direct sum of the cyclic p-subgroups $A_1, A_2, \ldots, A_m$ and is also the direct sum of the cyclic p-subgroups $B_1, B_2, \ldots, B_n$ then $m = n$ and, with a suitable renumbering if necessary, A_i is isomorphic to B_i $(i = 1, 2, \ldots, n)$.

Proof

The Corollary above yields the first statement of the theorem. We now prove the second statement. First of all, let A be an elementary Abelian p-group. Then all non-zero elements of A have order p and so any cyclic subgroup of A has order p.

Thus since A is the direct sum of cyclic subgroups $A_1, A_2, \ldots, A_m$ and of cyclic subgroups $B_1, B_2, \ldots, B_n$ each of these subgroups has order p and so

$$p^m = |A_1 \times A_2 \times \ldots \times A_m| = |A| = |B_1 \times B_2 \times \ldots \times B_n| = p^n.$$

Thus $m = n$ and the result holds.

We shall argue by induction and we make the assumption that the result is true for all Abelian p-groups of orders strictly less than $|A|$.

Let now A be not an elementary Abelian p-group. Suppose the notation is chosen so that $A_1, A_2, \ldots, A_r$ are cyclic groups of orders $\geq p^2$ and $A_{r+1}, A_{r+2}, \ldots, A_m$ are cyclic p-groups of orders p $(1 < r \leq m)$. Similarly suppose the notation is chosen so that $B_1, B_2, \ldots, B_s$ are cyclic subgroups of orders $\geq p^2$ and $B_{s+1}, B_{s+2}, \ldots, B_n$ are cyclic p-groups of orders p $(1 < s \leq n)$. Letting pA be the subgroup $\{pa \mid a \in A\}$ (that this is a p-subgroup is easily verified) we have, in an obvious adaptation of this notation,

$$pA = pA_1 \times pA_2 \times \ldots \times pA_r \times pA_{r+1} \times \ldots \times pA_m$$
$$= pA_1 \times pA_2 \times \ldots \times pA_r,$$

and

$$pA = pB_1 \times pB_2 \times \ldots \times pB_s \times pB_{s+1} \times \ldots \times pB_n$$
$$= pB_1 \times pB_2 \times \ldots \times pB_s.$$

By induction $r = s$ and, with a suitable renumbering if necessary, pA_i is isomorphic to pB_i $(i = 1, 2, \ldots, r)$. Hence A_i is isomorphic to B_i $(i = 1, 2, \ldots, r)$. Thus the direct products $A_1 \times A_2 \times \ldots \times A_r$ and $B_1 \times B_2 \times \ldots \times B_r$ are isomorphic.

But then

$$|A_{r+1} \times A_{r+2} \times \ldots \times A_m| = |A : A_1 \times A_2 \times \ldots \times A_r|$$
$$= |A : B_1 \times B_2 \times \ldots \times B_r|$$
$$= |B_{r+1} \times B_{r+2} \times \ldots \times B_n|.$$

Hence $p^{m-r} = p^{n-r}$ and so $m = n$ and this completes the proof. $\qquad\square$

Exercises 6.4

1. Let A be an Abelian group. Let M be the set of elements of A of finite order. Prove that M is a subgroup of A and that A/M contains no non-trivial elements of finite order.

2. Let A be an Abelian group. Let $n \in N$. Prove that $nA = \{na \mid a \in A\}$ is a subgroup of A.

3. Up to isomorphism find all Abelian groups of order 20.

4. Up to isomorphism find all Abelian groups of order 72.

5. Up to isomorphism find all Abelian groups of order 2250.

6.5 p-Groups and Sylow Subgroups

From the Fundamental Theorem of Abelian Groups every Abelian group of order $p_1^{\alpha_1} p_2^{\alpha_2} \ldots p_n^{\alpha_n}$, where $p_1, p_2, \ldots, p_n$ are distinct primes and $\alpha_i > 0$ $(i = 1, 2, \ldots, n)$, has a subgroup of order $p_i^{\alpha_i}$ $(i = 1, 2, \ldots, n)$. We have here a number-theoretic relationship between the order of an Abelian group and of the orders of certain subgroups of the group. We develop this relationship for non-Abelian groups but first we must investigate p-groups.

We begin with a useful counting result.

Theorem 10

Let G be a finite group. Let $Z(G)$ be the centre of G and let $C_1, C_2, \ldots, C_N$ be the conjugacy classes of G, each of which contains at least two elements of G. Let $x_r \in C_r$ and let $C_G(x_r)$ be the centralizer of x_r in G $(r = 1, 2, \ldots, N)$. Then

$$|G| = |Z(G)| + \sum_{r=1}^{N} |G : C_G(x_r)|.$$

Proof

$Z(G)$ consists of those elements of G which are self-conjugate. Thus

$$G = Z(G) \cup C_1 \cup C_2 \cup \ldots \cup C_N$$

is a disjoint union. By the Corollary to Theorem 20, Chapter 4, the number of elements in C_r is given by $|G : C_G(x_r)|$ and so, from the disjoint union above,

$$|G| = |Z(G)| + \sum_{r=1}^{N} |G : C_G(x_r)|. \qquad \square$$

Theorem 11

Let G be a finite group of order p^n $(n \geq 1)$ where p is a prime. Then the centre $Z(G)$ of G is non-trivial.

Proof

In the notation of Theorem 10, we have $|G : C_G(x_r)| > 1$ and, necessarily, $|G : C_G(x_r)|$ divides $|G|$. Thus since $|G| = p^n$ we deduce that p divides $|G : C_G(x_r)|$, $r = 1, 2, \ldots, N$. Hence p divides $\sum_{r=1}^{N} |G : C_G(x_r)|$.

Thus as

$$p^n = |G| = |Z(G)| + \sum_{r=1}^{N} |G : C_G(x_r)|$$

we conclude that p divides $|Z(G)|$ and so we obtain the result.   □

We give some examples showing the importance of Theorem 11.

Examples 20

1. Let G be a group of order p^2 where p is a prime. Then we may prove that G is Abelian as follows. Certainly we now know that the centre $Z(G)$ of G is non-trivial and so must be of order p or p^2. But if $Z(G)$ is of order p then $G/Z(G)$ is also of order p and so is cyclic. By Example 25 no. 3 of Section 4.4 G is then Abelian.

   A group of order p^3 need not be Abelian. Both the quaternion and dihedral groups of order 8 are non-Abelian.

2. Let G be a group of order p^n $(n \geq 1)$ where p is a prime. Then we assert that G contains a normal subgroup of index p. We know that $Z(G) \neq \{e\}$. If $Z(G) = G$ then G is Abelian and the reader should verify that the assertion follows from the Fundamental Theorem of Abelian Groups. We suppose therefore that $Z(G) \neq G$ and we intend to argue by induction on $|G|$.

   Thus by the induction assumption $G/Z(G)$ has a normal subgroup of index p. By Exercises 5.4, no. 4, the normal subgroup of index p may be written as $H/Z(G)$ where H is a normal subgroup of G. Then $|G/Z(G) : H/Z(G)| = p$ implies that

$$\frac{|G|}{|Z(G)|} \times \frac{|Z(G)|}{|H|} = p$$

   and so $|G : H| = \dfrac{|G|}{|H|} = p$, proving the assertion.

We come now to the first of two major and fascinating theorems on the structure of finite groups. Insofar as this text is concerned they will represent the culmination of our work in group theory. A.-L. Cauchy (1789–1857) proved that every group of order $p^a m$, where p is a prime not dividing m, contains a subgroup of order p, but it was P.L. Sylow (1832–1918) who established the existence of a subgroup of order p^a and which was subsequently named after him.

Theorem 12 (Sylow)

Let G be a finite group of order $p^a m$ where p is a prime not dividing m. Then G has a subgroup of order p^a.

Proof

If G is Abelian the result has already been shown. Suppose G is non-Abelian and make the induction assumption that the result is true for all groups of orders strictly less than $|G|$.

In the notation of Theorem 10 we have

$$p^a m = |G| = |Z(G)| + \sum_{r=1}^{N} |G : C_G(x_r)|.$$

We consider two cases. If p divides $|G : C_G(x_r)|$ for $r = 1, 2, \ldots, N$ then p divides $|Z(G)|$. But then $Z(G)$ has a subgroup Q of order p^b where $1 \leq b \leq a$. Since Q is a subgroup of the centre of G, Q is normal in G and so we may form G/Q which has order $p^{a-b} m$. By the induction assumption G/Q has a subgroup of order p^{a-b}. We let P/Q be this subgroup. Then $|P/Q| = p^{a-b}$ and so $|P| = |P/Q||Q| = p^{a-b} p^b = p^a$, proving the result.

On the other hand, if p does not divide $|G : C_G(x_r)|$ for all $r \in \{1, 2, \ldots, N\}$ then we may suppose that p does not divide $|G : C_G(x_i)|$, say. Then $|C_G(x_i)| = p^a n$ for some proper divisor n of m. By the induction assumption $C_G(x_i)$ has a subgroup of order p^a. The proof is now complete.   □

Definition 18

Let G be a finite group of order $p^a m$ where p is a prime and p does not divide m. Then a subgroup of G of order p^a is called a **Sylow p-subgroup** of G.

We note that a Sylow p-group of G has the greatest possible order of all p-subgroups of G.

Examples 21

1. A group G of order $8897850 = 2.3^4.5^2.13^3$ has Sylow subgroups corresponding to the primes 2, 3, 5 and 13. These Sylow subgroups have orders 2, 81, 25 and 2197 respectively.

2. Let us examine the Sylow subgroups of the symmetric group S_4. The order of the S_4 is $4! = 4.3.2.1 = 24 = 2^3.3$. The subgroup $V = \{(1), (1\ \ 2)(3\ \ 4), (1\ \ 3)(2\ \ 4), (1\ \ 4)(2\ \ 3)\}$ is a normal subgroup of order 4, and is a Klein four-group. V is contained in every Sylow 2-subgroup. There are, in fact,

exactly three Sylow 2-subgroups which are

$$V \cup V(1 \quad 2), V \cup V(1 \quad 3) \text{ and } V \cup V(1 \quad 4).$$

One Sylow 3-subgroup is $P = \{(1), (1 \quad 2 \quad 3), (1 \quad 3 \quad 2)\}$. The normalizer of P, $N_G(P)$, is given by $N_G(P) = P \cup P(1 \quad 2) = \{(1), (1 \quad 2 \quad 3), (1 \quad 3 \quad 2),$ $(1 \quad 2), (1 \quad 3), (2 \quad 3)\}$.

We obtain a coset decomposition

$$S_4 = N_G(P) \cup N_G(P)(1 \quad 4) \cup N_G(P)(2 \quad 4) \cup N_G(P)(3 \quad 4).$$

There are exactly four Sylow 3-subgroups which are

$$P, (1 \quad 4)^{-1}P(1 \quad 4), (2 \quad 4)^{-1}P(2 \quad 4), (3 \quad 4)^{-1}P(3 \quad 4).$$

Lemma 8

Let P be a normal Sylow p-subgroup of the finite group G. Let Q be a p-subgroup of G. Then $Q \subseteq P$.

Proof

Since P is normal in G, PQ is a subgroup of G. Since PQ/P is isomorphic to $Q/(P \cap Q)$ it follows that $|PQ/P| = |Q/(P \cap Q)|$ and so $|PQ/P|$ is a power of p. Thus PQ is a p-subgroup of G. But $P \subseteq PQ$ and P is a Sylow p-subgroup of G. Thus $P = PQ$ and so $Q \subseteq P$. □

Corollary

Let P be a Sylow p-subgroup of the finite group G. Let Q be a p-subgroup of the normalizer $N_G(P)$ of P. Then $Q \subseteq P$.

Proof

P is a normal Sylow p-subgroup of $N_G(P)$. Since $Q \subseteq N_G(P)$ we deduce, from Lemma 8, that $Q \subseteq P$. □

We have now reached the second of Sylow's famous theorems.

Theorem 13 (Sylow)

Let G be a finite group of order $p^a m$ where p is a prime and p does not divide m. Then the Sylow p-subgroups of G form a complete set of conjugate subgroups of G and if their number is n then $n = |G : N_G(P)|$ and $n \equiv 1 \pmod{p}$.

Proof

Let P be a Sylow p-subgroup of G and let $\{P_1, P_2, \ldots, P_n\}$ be a complete set of conjugate Sylow p-subgroups of G. Part of what we are required to show is that, of necessity, $P \in \{P_1, P_2, \ldots, P_n\}$.

Now, evidently, P acts as a permutation group on $\{P_1, P_2, \ldots, P_n\}$ if we define for $x \in P$, $x.P_i = xP_ix^{-1}$ $(i = 1, 2, \ldots, n)$. By Theorem 6, this action defines an equivalence relation $\sim$ on $\{P_1, P_2, \ldots, P_n\}$. Consider P_i. We want to calculate the number of $P_1, P_2, \ldots, P_n$ in the same equivalence class of P_i. In other words we want to know the number of elements in the orbit containing P_i. But, by Theorem 6, this number is $|P : \mathrm{Stab}_P(P_i)|$ where $\mathrm{Stab}_P(P_i)$ is the stabilizer of P_i under the action of P. We have

$$\mathrm{Stab}_P(P_i) = \{x \in P \,|\, x.P_i = P_i\} = \{x \in P \,|\, xP_ix^{-1} = P_i\}$$
$$= \{x \in P \,|\, x \in N_G(P_i)\} = P \cap N_G(P_i),$$

where $N_G(P_i)$ is the normalizer of P_i in G. But $|P : \mathrm{Stab}_P(P_i)|$ is a power of p and hence $|P : \mathrm{Stab}_P(P_i)| = 1$ if and only if $P = \mathrm{Stab}_P(P_i) = P \cap N_G(P_i)$. But this is possible if and only if $P \subseteq N_G(P_i)$ which, according to Lemma 8, would imply that $P \subseteq P_i$ and so $P = P_i$.

We may draw two conclusions from this argument. If we have $P \notin \{P_1, P_2, \ldots, P_n\}$ then for no i is $|P : \mathrm{Stab}_P(P_i)| = 1$ and so the number of elements in each orbit is divisible by p and hence $n \equiv 0 \pmod{p}$. On the other hand we may certainly choose P from $\{P_1, P_2, \ldots, P_n\}$, say we choose $P = P_1$. Then in this case the number in the orbit containing P_1 is precisely 1 but the numbers of elements in each of the remaining orbits is divisible by p and thus $n \equiv 1 \pmod{p}$. Thus if we are able to choose P not in $\{P_1, P_2, \ldots, P_n\}$, we have $n \equiv 0 \pmod{p}$ and $n \equiv 1 \pmod{p}$. Thus we have a contradiction which is only resolved if we cannot choose P not in $\{P_1, P_2, \ldots, P_n\}$. It follows therefore that the set $\{P_1, P_2, \ldots, P_n\}$ is the complete set of Sylow p-subgroups and these are therefore conjugate to one another. Furthermore we have $n \equiv 1 \pmod{p}$.

That $n = |G : N_G(P)|$ is obvious and the result is proved. $\qquad\square$

Examples 22

1. Let us prove that a group G of order 175 is necessarily Abelian. From the factorization, $175 = 5^2.7$, the group G has Sylow 5-subgroups of order 25 and Sylow 7-subgroups of order 7.

   Suppose there are m Sylow 5-subgroups and n Sylow 7-subgroups. Then m divides $5^2.7$ and $m = 1 + 5r$ for some $r \in \{0, 1, 2, \ldots\}$. If $r > 0$ then m must divide 7 which is impossible. Thus $r = 0$ and $m = 1$. Also n divides $5^2.7$ and $n = 1 + 7s$ for some $s \in \{0, 1, 2, \ldots\}$. If $s > 0$ then n must divide 5^2

which is impossible. Hence the Sylow 5-subgroup P and the Sylow 7-subgroup Q are both unique. Thus G is isomorphic to the direct product $P \times Q$ and is therefore Abelian as P and Q are Abelian.

2. Let G be a finite group and let P be a Sylow p-subgroup of G. Let H be a subgroup of G such that $N_G(P) \subseteq H$. We assert that $N_G(H) = H$.

The proof of the assertion exemplifies a common trick in regard to using Sylow subgroups. Let $x \in G$ be such that $x^{-1}Hx = H$. Then we wish to prove that $x \in H$. Certainly

$$x^{-1}Px \subseteq x^{-1}N_G(P)x \subseteq x^{-1}Hx = H.$$

Then P and $x^{-1}Px$ are both Sylow p-subgroups of H. Then the trick is to invoke Theorem 13 and to observe that P and $x^{-1}Px$ must be conjugate in H (and not merely in G). Thus we have $h \in H$ such that $h^{-1}Ph = x^{-1}Px$. But then

$$P = hx^{-1}Pxh^{-1} = (xh^{-1})^{-1}P(xh^{-1})$$

and so $xh^{-1} \in N_G(P) \subseteq H$ from which $x \in Hh = H$, completing the proof.

Exercises 6.5

1. Let G be a p-group. Let H be a proper subgroup of G. Prove that H is not equal to its normalizer in G. Deduce that any subgroup of G of index p is normal in G.

2. Prove that groups of orders 45 and 207 are Abelian.

3. Let P be a Sylow p-subgroup of a finite group G. Let N be a normal subgroup of G. Prove that $P \cap N$ is a Sylow p-subgroup of N and that PN/N is a Sylow p-subgroup of G/N. (Hint: use the Second Isomorphism Theorem.)

4. Prove that a group of order pq where p and q are primes has a proper normal subgroup.

5. Let G be a group of order 60. How many Sylow 5-subgroups may such a group have?

6. Prove that a group of order 351 has either a normal Sylow 3-subgroup or a normal Sylow 13-subgroup.

7. Prove that the Sylow 2-subgroups of the symmetric group S_4 are dihedral.

Hints to Solutions

1.1

1. $A \cup B = \{u, v, w, x, y\}$, $A \cap B = \{w\}$, $A \cap C = \emptyset$, $A \setminus B = \{u, v\}$, $B \setminus A = \{x, y\}$, $A' = \{x, y, z\}$, $A \setminus (B \cup C) = \{u, v\}$, $(A \cap B) \cup (B \setminus A) = \{w, x, y\}$.

2. $(X \cap Y)' = \{a, c, 1, 2, 3\} = X' \cup Y'$, $(X \cup Z)' = \emptyset = X' \cap Z'$.

3. $A \cup X = \{2, 4\}$ implies $1, 5 \notin X$. $A \cup X = \{1, 2, 4, 5, 6\}$ implies $6 \in X$. $X = \{2, 4, 6\}$.

4. $A \subseteq B$ implies $B \subseteq A \cup B \subseteq B \cup B = B$.

5. No rule is given. A could be either or neither.

6. k_i choices for elements of A_i $(i = 1, 2, \ldots, n)$.

1.2

1. (i)

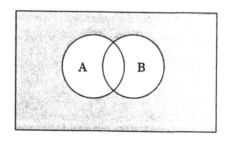

Shaded area represents $(A \cup B)'$ and $A' \cap B'$.

(ii) $x \in (A \cup B)'$ implies $x \notin A \cup B$ and so $x \notin A$, $x \notin B$. Thus $x \in A'$, $x \in B'$ and $x \in A' \cap B'$. $y \in A' \cap B'$ implies $y \in A'$, $y \in B'$. Thus $y \notin A$, $y \notin B$ and so $y \notin A \cup B$ and $y \in (A \cup B)'$.

2. (i)

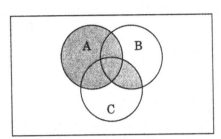

Shaded area represents $A \cup (B \cap C)$ and $(A \cup B) \cap (A \cup C)$.

(ii) $x \in A \cup (B \cap C)$ implies that $x \in A$ or $x \in B \cap C$. But $x \in A$ implies that $x \in (A \cup B) \cap (A \cup C)$. $x \in B \cap C$ implies $x \in B$, $x \in C$ and so it follows that $x \in (A \cup B) \cap (A \cup C)$. $y \in (A \cup B) \cap (A \cup C)$ implies $y \in A \cup B$ and $y \in A \cup C$. If $y \in A$ then $y \in A \cup (B \cap C)$. If $y \notin A$ then $y \in B$ and $y \in C$ and so $y \in B \cap C$. Thus $y \in A \cup (B \cap C)$.

3. No. Try $A = \{1,2\}, B = C = \{1\}$. If $A \subseteq C$ then
$$A \cup (B \cap C) = (A \cup B) \cap (A \cup C) = (A \cup B) \cap C.$$
If $(A \cup B) \cap (A \cup C) = (A \cup B) \cap C$ then
$$A = A \cap A \subseteq (A \cup B) \cap (A \cup C) = (A \cup B) \cap C \subseteq C.$$

4. Draw a Venn diagram. Let E, B, S be the sets of readers of the *Economist*, the *Bangkok Post* and the *Straits Times* respectively. $|E \cup B \cup S| = |E| + |B| + |S| - |E \cap B| - |E \cap S| - |B \cap C| + |E \cap B \cap S| = 250 + 411 + 315 - 72 - 51 - 31 + 1 = 823$. $1000 - 823 = 177$ do not read any of the publications.

5. Draw a Venn diagram. Let I, S, L be the sets of children who like ice cream, sweets and fizzy lemonade respectively. $|I \cap S \cap L| = 31$, $|I \cap L| = 36$, $|(I \cap L) \setminus (I \cap S \cap L)| = 5$, $|I \cap S'| = 7$, $|I \cap (S \cup L)'| = 2$, $|S \cap L| = 80$, $|(S \cap L) \setminus (I \cap S \cap L)| = 49$, $|L \cap (I \cup S)'| = 149 - 5 - 31 - 49 = 64$, $|S| = 110$. Hence $|I \cup S \cup L| = 110 + 2 + 5 + 64 = 181$. $200 - 181 = 19$ dislike all three items.

1.3

1. $(g \circ f)(a) = 0, (g \circ f)(b) = (g \circ f)(c) = 1$.

2. Only $q \circ p$ is defined.

3. $(f \circ g)(n) = 6n + 10$, $(g \circ f)(n) = 6n + 5$, $(f \circ h)(n) = -12n$, $(h \circ f)n = -12n$, $(g \circ h)(n) = -18n + 5$, $(h \circ g)(n) = -18n - 30$.

4. (i) If $g(f(a)) = g(f(a'))$ $(a, a' \in A)$ then $f(a) = f(a')$ and then $a = a'$.

   (ii) If $f(a) = f(a')$ then $g(f(a)) = g(f(a'))$ and so $a = a'$.

   (iii) Let $c \in C$. If there exists $a \in A$ such that $(g \circ f)(a) = c$, then $g(f(a)) = c$. But if $A = C = \{1, 2\}$, $B = \{1, 2, 3\}$ and $f(1) = 1$, $f(2) = 2$, $g(1) = 1$, $g(2) = g(3) = 2$ then $g \circ f$ is bijective, g is not injective and f is not surjective.

5. (i) If $x \in f(A)$, $x = f(a)$ $(a \in A)$, $A \subseteq B$ and so $x \in f(B)$.

   (ii) $A \subseteq A \cup B$ implies $f(A) \cup f(B) \subseteq f(A \cup B)$. If $x \in f(A \cup B)$, then $x = f(u)$ $(u \in A \cup B)$. If $u \in A$ then $f(u) \in f(A)$ and if $u \in B$ then $f(u) \in f(B)$. Thus $u \in f(A) \cup f(B)$.

   (iii) $A \cap B \subseteq A$, etc.

6. $n^2 - 3n + 5 = 3$ if and only if $n = 1, 2$. $f(1) = f(2) = 3$. $A \cap B = \varnothing$ and $f(A \cap B) = \varnothing$.

1.4

1. $a^2 = a^2$. $a^2 = b^2$ implies $b^2 = a^2$. $a^2 = b^2$, $b^2 = c^2$ implies $a^2 = c^2$.

2. $a \sim a$ since $a - a = 0$. $a \sim b$, $a - b$ even, $b - a$ even, $b \sim a$. $a \sim b$, $b \sim c$, $a - b$, $b - c$ even, $a - c = (a - b) + (b - c)$, $a - c$ even, $a \sim c$.

3. $a \sim a$ since 6 divides $0 = a - a$, etc.

4. For each $q \in \mathbb{Q}$, equivalence class $\{q + x \mid x \in \mathbb{Z}\}$.

5. (i) Yes. Equivalence class is straight line through 0.

   (ii) Yes. Equivalence class is circumference of circle, centre at 0.

6. Recall that an equivalence relation corresponds to a partition.

   $X = \{x\}$: $\{x\}$. One relation.

   $X = \{x, y\}$: $\{x\}, \{y\}$; $\{x, y\}$. Two relations.

   $X = \{x, y, z\}$: $\{x\}, \{y\}, \{z\}$; $\{x, y\}, \{z\}$; $\{x, z\}, \{y\}$; $\{y, z\}, \{x\}$; $\{x, y, z\}$. Five relations.

$X = \{x, y, z, t\}$: $\{x\}$, $\{y\}$, $\{z\}$, $\{t\}$; $\{x, y\}$, $\{z\}$, $\{t\}$ and five other similar partitions; $\{x, y\}$, $\{z, t\}$ and two other similar partitions; $\{x, y, z\}$, $\{t\}$ and three other similar partitions. Fourteen relations.

7. No. 5 divides $14 - 9$, 7 divides $9 - 2$ but neither 5 nor 7 divides $14 - 2$.

8. ρ is the equivalence relation corresponding to the partition $\{a, b, c\}$, $\{x\}$, $\{y\}$, $\{z\}$. σ is not an equivalence relation since $x\sigma a$, $a\sigma y$ but $x\sigma y$ is false.

9. We do not necessarily have $a \sim a$ for all $a \in S$.

1.5

1. $1(1 + 1) = \dfrac{1}{3}1(1 + 1)(1 + 2)$.

$$\sum_{r=1}^{n+1} r(r + 1) = \sum_{r=1}^{n} r(r + 1) + (n + 1)(n + 2)$$

$$= \frac{1}{3}n(n + 1)(n + 2) + (n + 1)(n + 2)$$

$$= \frac{1}{3}n(n + 1)(n + 2)(n + 3)$$

2. $1^3 = \dfrac{1}{4}1^2(1 + 1)^2$.

$$\sum_{r=1}^{n+1} r^3 = \frac{1}{4}n^2(n + 1)^2 + (n + 1)^3 = \frac{1}{4}(n + 1)^2(n + 1)^2.$$

3. $a \neq 1$, $\quad a^0 = \dfrac{1 - a}{1 - a}$.

$$\sum_{r=1}^{n+1} a^{r-1} = \frac{1 - a^n}{1 - a} + a^n = \frac{1 - a^{n+1}}{1 - a}.$$

4. $\dfrac{1}{1(1 + 1)} = \dfrac{1}{1 + 1}$.

$$\sum_{r=1}^{n+1} \frac{1}{r(r + 1)} = \frac{n}{n + 1} + \frac{1}{(n + 1)(n + 2)} = \frac{n + 1}{n + 2}.$$

5. $(5 - 2) = \dfrac{1}{2}(5 + 1)$.

$$\sum_{r=1}^{n+1} (5r - 2) = \frac{n}{2}(5n + 1) + [5(n + 1) - 2] = \frac{1}{2}(n + 1)[5(n + 1) + 1].$$

6. $2 = 2$.

$$\sum_{r=1}^{n+1} r2^r = 2 + (n-1)2^{n+1} + (n+1)2^{n+1} = 2 + n2^{n+2}.$$

7. $1 + x \geq 1 + x$.

$(1+x)^{n+1} \geq (1+nx)(1+x) = 1 + (n+1)x + x^2 \geq 1 + (n+1)x$.

8. $3(6.2 + 1) = 39 \leq 2^6$. Then $n \geq 6$,

$3[2(n+1) + 1] = 3(2n+1) + 3.2 \leq 2^n + 3.2 \leq 2^n + 2^n = 2^{n+1}$.

9. $3.8^2 = 3.2^6 \leq 2^2.2^6 = 2^8$. Then $n \geq 8$,

$3(n+1)^2 = 3n^2 + 6n + 3 \leq 2^n + 3(2n+1) \leq 2^n + 2^n = 2^{n+1}$.

10. $\sqrt{2} \leq \dfrac{3}{2}$ implies $2(\sqrt{2} - 1) \leq \dfrac{1}{\sqrt{1}}$.

$$\sum_{r=1}^{n+1} \frac{1}{\sqrt{r}} \geq 2(\sqrt{n+1} - 1) + \frac{1}{\sqrt{n+1}}.$$

But $2(\sqrt{n+1} - 1) + \dfrac{1}{\sqrt{n+1}} - 2(\sqrt{n+2} - 1)$

$= \dfrac{1}{\sqrt{n+1}} - 2(\sqrt{n+2} - \sqrt{n+1}) = \dfrac{1}{\sqrt{n+1}} = 2\left[\dfrac{(n+2) - (n+1)}{\sqrt{n+2} + \sqrt{n+1}}\right]$

$= \dfrac{1}{\sqrt{n+1}} - \dfrac{2}{\sqrt{n+2} + \sqrt{n+1}} \geq 0$. Thus $\displaystyle\sum_{r=1}^{n+1} \frac{1}{\sqrt{r}} \geq 2(\sqrt{n+2} - 1)$.

11. 1, 2, 3, 5, 8, 13, 21, 34, 55.

(i) $u_1 = 1 \leq \dfrac{7}{4}$.

$u_{n+1} = u_n + u_{n-1} \leq \left(\dfrac{7}{4}\right)^n + \left(\dfrac{7}{4}\right)^{n-1}$

$= \left(\dfrac{7}{4}\right)^{n-1}\left(\dfrac{7}{4} + 1\right) \leq \left(\dfrac{7}{4}\right)^{n-1}\left(\dfrac{7}{4}\right)^2 \leq \left(\dfrac{7}{4}\right)^{n+1}$

(ii) $\quad 1 = \dfrac{1}{\sqrt{5}} \left[\left(\dfrac{1+\sqrt{5}}{2} \right)^2 - \left(\dfrac{1-\sqrt{5}}{2} \right)^2 \right]$

$$u_{n+1} = u_n + u_{n-1} = \dfrac{1}{\sqrt{5}} \left(\dfrac{1+\sqrt{5}}{2} \right)^{n+1} - \dfrac{1}{\sqrt{5}} \left(\dfrac{1-\sqrt{5}}{2} \right)^{n-1}$$

$$+ \dfrac{1}{\sqrt{5}} \left(\dfrac{1+\sqrt{5}}{2} \right)^{n} - \dfrac{1}{\sqrt{5}} \left(\dfrac{1-\sqrt{5}}{2} \right)^{n}$$

$$= \dfrac{1}{\sqrt{5}} \left(\dfrac{1+\sqrt{5}}{2} \right)^{n} \left[\dfrac{1+\sqrt{5}}{2} + 1 \right] - \dfrac{1}{\sqrt{5}} \left(\dfrac{1-\sqrt{5}}{2} \right)^{n} \left[\dfrac{1-\sqrt{5}}{2} + 1 \right]$$

$$= \dfrac{1}{\sqrt{5}} \left(\dfrac{1+\sqrt{5}}{2} \right)^{n} \left[\dfrac{3+\sqrt{5}}{2} \right] - \dfrac{1}{\sqrt{5}} \left(\dfrac{1-\sqrt{5}}{2} \right)^{n} \left[\dfrac{3-\sqrt{5}}{2} \right]$$

$$= \dfrac{1}{\sqrt{5}} \left(\dfrac{1+\sqrt{5}}{2} \right)^{n} \left(\dfrac{1+\sqrt{5}}{2} \right)^{2} - \dfrac{1}{\sqrt{5}} \left(\dfrac{1-\sqrt{5}}{2} \right)^{n} \left(\dfrac{1-\sqrt{5}}{2} \right)^{2}$$

$$= \dfrac{1}{\sqrt{5}} \left(\dfrac{1+\sqrt{5}}{2} \right)^{n+2} - \dfrac{1}{\sqrt{5}} \left(\dfrac{1-\sqrt{5}}{2} \right)^{n+2}.$$

1.6

1. $\mathbf{N}$, $\mathbf{Z}$ countable. Hence result.

2. Identify $a + b\sqrt{2}$ with (a, b); $\mathbf{Z} \times \mathbf{Z}$ is countable.

3. $A = (A \cup B) \cup (A \setminus B)$ is a disjoint union. Hence result.

4. $\{0, 1\}^{\mathbf{N}}$ uncountable. Hence $\mathbf{N}^{\mathbf{N}}$ uncountable.

5. Consider array $x_1, x_2, \ldots, x_{1q}$

$$x_{21}, x_{22}, \ldots, x_{2q}$$

$$\cdots$$

2.1

1. (i)   1, 2, 3, 4, 6, 8, 9, 12, 18, 24, 36, 72.

   (ii) 1, 2, 3, 4, 6, 7, 12, 14, 21, 28, 42, 84.

   (iii) 1, 23, 29, 667.

2. If b divides $-a$ then $b = (-a)c = -ac$ for some $c \in \mathbf{Z}$, etc.

3. (i) $2^3.3$, (ii) 3.31, (iii) $5^2.7^2$, (iv) 17^3, (v) $2^4.3^4.11$.

2.2

1. (i) 3, (ii) 12, (iii) 1, (iv) 15.

2.3

1. (i) $q = 1, r = 2.$
   (ii) $q = 3, r = 3.$
   (iii) $q = 6, r = 31.$
   (iv) $2513 = 46.54 + 29, -2513 = 46(-54) - 29 = 46(-55) + (46 - 29),$
        $q = -55, r = 46 - 29 = 17.$
   (v) $q = -71, r = 290.$
   (vi) $-52148 = (-732)71 - 176 = (-732).72 + (732 - 176), q = 72,$
        $r = 732 - 176 = 556.$

2. (i) $|xy|^2 = (xy)^2 = x^2 y^2 = [|x||y|]^2.$ Hence $|xy| = |x||y|.$
   (ii) $|x + y|^2 = (x + y)^2 = x^2 + y^2 + 2xy \leq |x|^2 + |y|^2 + 2|x||y| = (|x| + |y|)^2.$
        Hence $|x + y| \leq |x| + |y|.$
        $|1| - |-1| = 1 - 1 = 0, |1 - (-1)| = |2| = 2.$

2.4

1. $a = na', b = nb'(a', b' \in Z).$
   $ax + by = na'x + nb'y = n(a'x + b'y).$

2. (i) $44 = 32.1 + 12, 32 = 12.2 + 8, 12 = 8 + 4, 8 = 4.2.$ G.C.D. 4.
       $4 = 12 - 8 = 12 - (32 - 12.2) = 12.3 - 32 = 3(44 - 32) - 32 = 3.44 - 4.32.$
   (ii) $150 = 105 + 45, 105 = 45.2 + 15, 45 = 15.3.$ G.C.D. 15.
        $15 = 105 - 45.2 = 105 - (150 - 105)2 = 105.3 + 150(-2).$
   (iii) $3718 = 1222.3 + 52, 1222 = 52.23 + 26, 52 = 26.2.$ G.C.D. 26.
         $26 = 1222 - 52.23 = 1222 - (3718 - 1222.3).23 = 3718(-23) + 1222.70.$
   (iv) $96 = 0.764 + 96, 764 = 96.7 + 92, 96 = 92 + 4, 92 = 4.23.$ G.C.D. 4.
        $4 = 96 - 92 = 96 - (764 - 96.7) = 96.8 - 764.$
   (v) G.C.D. 2. $2 = 6214.1595 - 7224.1372.$
   (vi) G.C.D. 529. $529 = 152881.289 - 613640.72.$

3. $a^2 - b^2 = (a - b)(a + b), a^3 - b^3 = (a - b)(a^2 + b^2 + ab),$
   $a^3 + b^3 = (a + b)(a^2 + b^2 - ab).$ 3 divides $1 + 2$ but does not divide $1^2 + 2^2.$

2.5

1. Proof for $\sqrt{3}$ like proof for $\sqrt{2}$.

   Suppose $\sqrt{6} = \dfrac{m}{n}$ $(m, n \in \mathbb{N})$ where m, n have no common divisor $\neq 1$.

   $m^2 = 6n^2$ implies 2 divides m^2 and so m. Let $m = 2m'$. Then $4m'^2 = m^2 = 6n^2$ and so $2m'^2 = 3n^2$. Thus 2 divides $3n^2$ and so 2 divides n, etc.

2. $\sqrt{2} + \sqrt{3} = q \in \mathbb{Q}$ implies $\sqrt{6} = \dfrac{q^2 - 5}{2} \in \mathbb{Q}$, false.

3. $\pm\sqrt{2}$ irrational, $\sqrt{2} + (-\sqrt{2}) = 0$ rational.

4. $a, b, q \in \mathbb{Z}$, q prime to b. Then there exist $x, y \in \mathbb{Z}$ such that $1 = qx + by$ and $a = aqx + aby$.

5. Pairs 3, 5; 5, 7; 11, 13; 17, 19; 29, 31; 41, 43; 59, 61; 71, 73.

6. Apply sieve to numbers from 100 to 200.

7. $\pi(10) = 4$, $\pi(20) = 8$, $\pi(30) = 10$, $\pi(40) = 12$, $\pi(50) = 15$, $\pi(60) = 17$, $\pi(70) = 19$, $\pi(80) = 22$, $\pi(90) = 24$, $\pi(100) = 25$.

8. Eleven ways.

3.1

1. $x + 2$.

2. $x + \frac{2}{3}$.

3. $(x + 1)^2$.

4. $x^2 + 1$.

3.2

1. (i) $q(x) = 0$, $r(x) = x^3 + 3x^2 + 1$.
   (ii) $q(x) = x + 1$, $r(x) = -1$.
   (iii) $q(x) = \frac{2}{5}x + \frac{27}{25}$, $r(x) = \frac{52}{25}$.
   (iv) $q(x) = x^2 + x + 1$, $r(x) = -2$.
   (v) $q(x) = \frac{1}{2}x^2 + \frac{5}{4}$, $r(x) = -\frac{1}{4}$.

2. (i) $x^4 + x^3 + x + 1 = (x^2 + x + 1)(x^2 - 1) + (2x + 2)$,

   $$x^2 + x + 1 = (2x + 2)\left(\frac{x}{2}\right) + 1.$$

   G.C.D. 1. $1 = (x^2 + x + 1)\left(\frac{1}{2}x^3 - \frac{1}{2}x + 1\right) - \frac{x}{2}(x^4 + x^3 + x + 1)$.

(ii) $2x^3 + 10x^2 + 2x + 10 = 2(x^3 - 2x^2 + x - 2) + (14x^2 + 14)$,

$x^3 - 2x^2 + x - 2 = (14x^2 + 14)\left(\dfrac{1}{14}x - \dfrac{1}{7}\right)$.

G.C.D. $x^2 + 1$. $x^2 + 1 = \dfrac{1}{14}(2x^3 + 10x^2 + 2x + 10) - \dfrac{1}{7}(x^3 - 2x^2 + x - 2)$.

(iii) $x^4 - 4x^3 + 2x - 4 = (x^3 + 2)(x - 4) + 4$, $x^3 + 2 = 4\left[\dfrac{1}{2}(x^3 + 2)\right]$.

G.C.D. 1. $1 = \dfrac{1}{4}(x^4 - 4x^3 + 2x - 4) - \dfrac{1}{4}(x - 4)(x^3 + 2)$.

(iv) $x^3 + 5x^2 + 7x + 2 = (x^3 + 2x^2 - 2x - 1) + (3x^2 + 9x + 3)$,

$x^3 + 2x^2 - 2x - 1 = (3x^2 + 9x + 3)\left(\dfrac{1}{3}x - \dfrac{1}{3}\right)$.

G.C.D. $x^2 + 3x + 1$.

$x^2 + 3x + 1 = \dfrac{1}{3}(x^3 + 5x^2 + 7x + 2) - \dfrac{1}{3}(x^3 + 2x^2 - 2x - 1)$.

3.3

1. e, f identities. $e = ef = f$.

2. Check all axioms are satisfied.

3. R is closed under addition and multiplication and the other axioms are satisfied.

4. Note that $R = \mathbb{Z}_2$. Check axioms.

5. Let $x, y \in \mathbb{Z}$. Then $2x + 2y = 2(x + y)$, $(2x)(2y) = 2(2xy)$. Thus closure of $2\mathbb{Z}$ under addition and multiplication. Associativity of addition and multiplication, and also distributive laws, hold in $\mathbb{Z}$ and so in $2\mathbb{Z}$. Remaining axioms are clear. $2\mathbb{Z}$ has no identity element.

6. (i) If $a + b = c + b$ then

$$a = a + 0 = a + (b + (-b)) = (a + b) + (-b) = (c + b) + (-b)$$
$$= c + (b + (-b)) = c + 0 = c.$$

(ii) $-(ab) = -(ab) + 0 = -(ab) + 0b = -(ab) + (a + (-a))b$
$= (-ab) + (ab + (-a)b) = (-(ab) + ab) + (-a)b = 0 + (-a)b = (-a)b.$

7. If A, B are 2×2 matrices over $\mathbb{Q}$ then so also are $A + B$, AB. Matrix addition and multiplication are associative etc.

8. $R_1 \cap R_2 \neq \emptyset$ since $0 \in R_1 \cap R_2$. Let $x, y \in R_1 \cap R_2$. Then $x + y, -x, xy \in R_1$ and $x + y, -x, xy \in R_2$. Thus $x + y, -x, xy \in R_1 \cap R_2$ which is therefore a subring. The intersection of any collection of subrings is a subring.

4.1

1. For all $a, b, c \in \mathbb{Z}$, $a * b \in \mathbb{Z}$ and $a * (b * c) = a * (cb) = cb(a) = c(ba) = ba * c = (a * b) * c$, $a * 1 = 1a = a = a1 = 1 * a$.

2. $(f_{c,d} \circ f_{a,b})(x) = f_{c,d}(f_{a,b}(x)) = f_{c,d}(a + bx) = c + d(a + bx) = c + ad + bdx = f_{c+ad, bd}(x)$. Circle-composition is associative and $f_{0,1}$ is the identity for the monoid.

3. Verify associativity, $0 + a = a + 0 = a$.

4. There exists $a' \in S$ such that $aa' = e$ and $a'' \in S$ such that $a'a'' = e$. Then $ea = ea(e) = (ea)(a'a'') = ((ea)a')a'' = (e(aa'))a'' = (ee)a'' = ea'' = (aa')a'' = a(a'a'') = ae = a$. $a'' = ea'' = (aa')a'' = a(a'a'') = ae = a$.

5. Since S is finite there exists N such that every power of a is equal to one of $a, a^2, \ldots, a^N$. Let $p > 2N$. Then there exists q, $1 \leq q \leq N$, such that $a^p = a^q$ where $p > 2N \geq 2q$. Also $b^2 = (a^{p-q})^2 = a^{2p-2q} = a^p a^{p-2q} = a^q a^{p-2q} = a^{p-q} = b$.

4.2

1. No. $1 \in \mathbb{N}$, $-1 \notin \{0\} \cup \mathbb{N}$.

2. $x = ex = (x^{-1}x)x = x^{-1}(xx) = x^{-1}x = e$.

3. $(ab^2)^{-1}(c^2 a^{-1})^{-1}(c^2 b^2 d)(ad)^{-2}a = b^{-2}a^{-1}ac^{-2}c^2 b^2 dd^{-2}a^{-2}a = d^{-1}a^{-1}$. $(abc)^{-1}(ab)^2 d(d^{-1}b^{-1})^2 bdc = c^{-1}b^{-1}a^{-1}ababdd^{-1}b^{-1}d^{-1}b^{-1}bdc = c^{-1}ac$.

4. No. $a(bc) = aa = c \neq b = dc = (ab)c$.

5. $(ac)(x) = a(c(x)) = a(1 - x) = \dfrac{(1 - x) - 1}{1 - x} = \dfrac{-x}{1 - x} = \dfrac{x}{x - 1} = d(x)$.

   Thus $ac = d$, etc.

6. Use de Moivre's Theorem.

7. Use de Moivre's Theorem.

8. Let f, g be strictly monotonic. Then $0 \leq x_1 < x_2 \leq 1$ implies $0 \leq g(x_1) < g(x_2) \leq 1$ and so $0 \leq (f \circ g)(x_1) < (f \circ g)(x_2) \leq 1$. Thus $f \circ g$ is strictly monotonic. Circle-composition is associative and $\iota(x) = x$ gives identity. Inverse f^{-1} of f is given by $f^{-1}(x) = y$ if and only if $f(y) = x$. Prove f^{-1} is strictly monotonic!

9. $\begin{pmatrix} a & b \\ -b & a \end{pmatrix}\begin{pmatrix} x & y \\ -y & x \end{pmatrix} = \begin{pmatrix} ax - by & ay + bx \\ -bx - ay & -by + ax \end{pmatrix} \in G.$ Matrix multiplication

is associative. $\begin{pmatrix} 1 & 0 \\ 0 & 1 \end{pmatrix} \in G,$ $\begin{pmatrix} a & b \\ -b & a \end{pmatrix}^{-1} = \dfrac{1}{a^2 + b^2}\begin{pmatrix} a & -b \\ b & a \end{pmatrix} \in G.$

G is Abelian.

10. (i) $c^2 - \left(\dfrac{u+v}{1 + \frac{uv}{c^2}}\right)^2 = \dfrac{(c^2 - u^2)(c^2 - v^2)}{c^2(1 + \frac{uv}{c})^2} \geq 0.$

(ii) Let $y = \dfrac{x - vt}{\sqrt{1 - \frac{v^2}{c^2}}},\ s = \dfrac{t - \frac{v}{c^2}x}{\sqrt{1 - \frac{v^2}{c^2}}}$

$(T_u \circ T_v)(x, t) = T_u(T_v(x, t)) = T_u(y, s)$

$= \left(\dfrac{y - us}{\sqrt{1 - \frac{u^2}{c^2}}}, \dfrac{s - \frac{u}{c^2}y}{\sqrt{1 - \frac{u^2}{c^2}}}\right) = \left(\dfrac{x - vt - u(t - \frac{v}{c^2}x)}{\sqrt{1 - \frac{u^2}{c^2}}\sqrt{1 - \frac{v^2}{c^2}}}, \dfrac{t - \frac{v}{c^2}x - \frac{u}{c^2}(x - vt)}{\sqrt{1 - \frac{u^2}{c^2}}\sqrt{1 - \frac{v^2}{c^2}}}\right)$

$= \left(\dfrac{(1 + \frac{uv}{c^2})x - (u + v)t}{\sqrt{1 - \frac{u^2}{c^2}}\sqrt{1 - \frac{v^2}{c^2}}}, \dfrac{(1 + \frac{uv}{c^2})t - \frac{(u+v)}{c^2}x}{\sqrt{1 - \frac{u^2}{c^2}}\sqrt{1 - \frac{v^2}{c^2}}}\right) = \left(\dfrac{x - wt}{X}, \dfrac{t - \frac{w}{c^2}x}{X}\right)$

where $X = \dfrac{\sqrt{1 - \frac{u^2}{c^2}}\sqrt{1 - \frac{v^2}{c^2}}}{1 + \frac{uv}{c^2}} = \sqrt{1 - \dfrac{w^2}{c^2}}.$

Circle-composition is associative, T_o is the identity and $(T_v)^{-1} = T_{-v}.$

4.3

1. Let $G = \{e, a, a^2, a^3, a^4, a^5\},\ a^6 = e.$ Cayley table is:

	e	a	a^2	a^3	a^4	a^5
e	e	a	a^2	a^3	a^4	a^5
a	a	a^2	a^3	a^4	a^5	e
a^2	a^2	a^3	a^4	a^5	e	a
a^3	a^3	a^4	a^5	e	a	a^2
a^4	a^4	a^5	e	a	a^2	a^3
a^5	a^5	e	a	a^2	a^3	a^4

From Cayley table $\{e, a\}$, $\{e, a^2, a^4\}$ are the proper subgroups.

2. Let $a, b \in A \cap B \cap C \cap \ldots$ Then $ab, a^{-1} \in A$, $ab, a^{-1} \in B, \ldots$ Thus $ab, a^{-1} \in A \cap B \cap C \ldots$

3. $\{e, a, b\}$ is a normal subgroup. $\{e, c\}$, $\{e, d\}$, $\{e, f\}$ are conjugate subgroups. No other proper subgroups.

4. $x \in G$, $a \in A \cap B \cap C \ldots$ Then $x^{-1}ax \in A$, $x^{-1}ax \in B$, etc.

5. Let $u, v \in a^{-1}Ha$. $u = a^{-1}ha$, $v = a^{-1}ka(h, k \in H)$.
$uv = a^{-1}haa^{-1}ka = a^{-1}hka \in a^{-1}Ha$ and $u^{-1} = a^{-1}h^{-1}a \in H$. Thus $a^{-1}Ha$ is a subgroup. Hence $\bigcap_{x \in G} x^{-1}Hx$ is a subgroup. Let $c = \bigcap_{x \in G} x^{-1}Hx$ Then $c \in x^{-1}Hx$ for all $x \in G$. Let $y \in G$. $y^{-1}cy \in y^{-1}x^{-1}Hxy = (xy)^{-1}H(xy)$ for all $x \in G$. As x runs over the elements of G so does xy and so $y^{-1}cy \in x^{-1}Hx$ for all $x \in G$. Thus

$$y^{-1}cy \in \bigcap_{x \in G} x^{-1}Hx$$

which is therefore normal.

6. Let A, B be non-singular matrices. Then AB, A^{-1} are non-singular as $(AB)^{-1} = B^{-1}A^{-1}$ and $(A^{-1})^{-1} = A$, etc.

$$\begin{pmatrix} 1 & a \\ 0 & 1 \end{pmatrix} \begin{pmatrix} 1 & b \\ 0 & 1 \end{pmatrix} = \begin{pmatrix} 1 & a+b \\ 0 & 1 \end{pmatrix}, \begin{pmatrix} 1 & a \\ 0 & 1 \end{pmatrix}^{-1} = \begin{pmatrix} 1 & -a \\ 0 & 1 \end{pmatrix}.$$

Thus H is a subgroup which is obviously Abelian.

7. Let $c \in C_G(A)$, $x \in N_G(A)$. We need to prove $x^{-1}cx \in C_G(A)$. Let $a \in A$. Then $xax^{-1} \in A$ and so $(x^{-1}cx)^{-1}a(x^{-1}cx) = x^{-1}c^{-1}xax^{-1}cx = x^{-1}xax^{-1}x = a$. Hence we have $x^{-1}cx \in C_G(A)$.

8. Suppose NH is a subgroup. Then by closure $HN \subseteq NH$. Let $x \in NH$. Then $x^{-1} \in NH$, $x^{-1} = nh$ $(n \in N, h \in H)$. Then $x = h^{-1}n^{-1} \in HN$ and so $NH \subseteq HN$. Thus $HN = NH$.

    Suppose now $NH = HN$. Let $x, y \in NH$, $x = nh$, $y = mk$ where $m, n \in N, h, k \in H$. Since $hm \in HN = NH$, let $hm = m'h'$ $(m' \in N, h' \in H)$. Then $xy = nhmk = nm'h'k \in NH$ and $x^{-1} = h^{-1}n^{-1} \in HN = NH$. Hence NH is a subgroup.

9. Each H_i contains $|H_i| - 1 = |H| - 1$ elements other than e.
Thus $H_1 \cup H_2 \cup \ldots \cup H_n$ contains at most $n(|H| - 1)$ elements other than e and so $|H_1 \cup H_2 \cup \ldots \cup H_n| \le n(|H| - 1) + 1$. If $G = H_1 \cup H_2 \cup \ldots \cup H_n$, then $|G| \le n(|H| - 1) + 1 \le |G : H|(|H| - 1) + 1 = |G| - |G : H| + 1$ which is impossible unless $|G : H| = 1$ and $H = G$.

4.4

1. Let $G = N_G(H)a_1 \cup N_G(H)a_2 \cup \ldots \cup N_G(H)a_n$ be a coset decomposition. Then the conjugate subgroups $a_1^{-1}Ha_1,\ a_2^{-1}Ha_2, \ldots, a_n^{-1}Ha_n$ are distinct and there are no other conjugate subgroups.

2. $x \in G$ implies $x^{-1} \in Ha_i$ (some i). Thus $x \in a_i^{-1}H$ and so $G = a_1^{-1}H \cup a_2^{-1}H \cup \ldots \cup a_n^{-1}H$. $a_i^{-1}H = a_j^{-1}H$ if and only if $a_i^{-1} = a_j^{-1}h^{-1}$ $(h \in H)$, that is $a_i = ha_j$ or $Ha_i = Ha_j$.

3. $12\mathbb{Z},\ 12\mathbb{Z} + 1,\ \ldots,\ 12\mathbb{Z} + 11$.

4.5

1. $(g \circ f)(xy) = g(f(xy)) = g(f(x)f(y)) = g(f(x))g(f(y)) = (g \circ f)(x)(g \circ f)(y)$ $(x, y \in G)$.

2. $aabb = f(a)f(b) = f(ab) = abab$. Cancelling a and b gives $ab = ba$.

3. $f(x) = x^{-1}a^{-1}xa \in Z(G)$. $f(x)f(y) = x^{-1}a^{-1}xay^{-1}a^{-1}ya = y^{-1}(x^{-1}a^{-1}xa)$ $a^{-1}ya = (xy)^{-1}a^{-1}(xy)a = f(xy)$ $(x, y \in G)$.

4. $H/N = \{Nh | h \in H\}$. Let $x, y \in H$. $(Nx)(Ny) = Nxy$, $(Nx)^{-1} = Nx^{-1}$. Thus H/N is a subgroup of G/N. Let W be a subgroup of G/N. Let H be the subset of G given by $\{x \in G | Nx \in W\}$. Then prove that H is a subgroup of G, certainly $H/N = W$. $x^{-1}Hx = H$ if and only if $(Nx)^{-1}(H/N)(Nx) = H/N$.

5. Proper subgroups are $\{e, a, b, c\}$, $\{e, a, d, f\}$, $\{e, a, g, h\}$ and centre $Z(G) = \{e, a\}$. Classes are $\{e\}$, $\{a\}$, $\{b, c\}$, $\{d, f\}$, $\{e, g\}$.
   $G = Z(G) \cup Z(G)b \cup Z(G)d \cup Z(G)g$. $G/Z(G)$ is a Klein four-group.

5.1

1.

$\mathbb{Z}_2$

Sum	$\bar{0}$	$\bar{1}$
$\bar{0}$	$\bar{0}$	$\bar{1}$
$\bar{1}$	$\bar{1}$	$\bar{0}$

Product	$\bar{0}$	$\bar{1}$
$\bar{0}$	$\bar{0}$	$\bar{0}$
$\bar{1}$	$\bar{0}$	$\bar{1}$

$\mathbb{Z}_3$

Sum	$\bar{0}$	$\bar{1}$	$\bar{2}$
$\bar{0}$	$\bar{0}$	$\bar{1}$	$\bar{2}$
$\bar{1}$	$\bar{1}$	$\bar{2}$	$\bar{0}$
$\bar{2}$	$\bar{2}$	$\bar{0}$	$\bar{1}$

Product	$\bar{0}$	$\bar{1}$	$\bar{2}$
$\bar{0}$	$\bar{0}$	$\bar{0}$	$\bar{0}$
$\bar{1}$	$\bar{0}$	$\bar{1}$	$\bar{2}$
$\bar{2}$	$\bar{0}$	$\bar{2}$	$\bar{1}$

Z_4

Sum	$\bar{0}$	$\bar{1}$	$\bar{2}$	$\bar{3}$
$\bar{0}$	$\bar{0}$	$\bar{1}$	$\bar{2}$	$\bar{3}$
$\bar{1}$	$\bar{1}$	$\bar{2}$	$\bar{3}$	$\bar{0}$
$\bar{2}$	$\bar{2}$	$\bar{3}$	$\bar{0}$	$\bar{1}$
$\bar{3}$	$\bar{3}$	$\bar{0}$	$\bar{1}$	$\bar{2}$

Product	$\bar{0}$	$\bar{1}$	$\bar{2}$	$\bar{3}$
$\bar{0}$	$\bar{0}$	$\bar{0}$	$\bar{0}$	$\bar{0}$
$\bar{1}$	$\bar{0}$	$\bar{1}$	$\bar{2}$	$\bar{3}$
$\bar{2}$	$\bar{0}$	$\bar{2}$	$\bar{0}$	$\bar{2}$
$\bar{3}$	$\bar{0}$	$\bar{3}$	$\bar{2}$	$\bar{1}$

2. $x = 6$, $y = 3$, $z = 9$.

3. $x = a_2 u_1 m_1 + a_1 u_2 m_2 = a_2(1 - u_2 m_2) + a_1 u_2 m_2$

$= a_2 - a_2 u_2 m_2 + a_1 u_2 m_2 \equiv a_2 \pmod{m_2}$. Similarly $x \equiv a_1 \pmod{m_1}$.

4. Ring axioms are easily verified. $(1_D, 0_E)(0_D, 1_E) = (0_D, 0_E)$ implies that proper divisors of zero exist.

5. $x(x - 1) = 0$. Hence $x = 0$ or $x - 1 = 0$.

6. Verify assertions.

5.2

1. $x^2 + x + \bar{2} = (x - \bar{3})^2$, $x^2 + \bar{2}x + \bar{4} = (x - \bar{1})(x - \bar{4})$, $x^2 + x + \bar{3}$ does not factorize, $x^6 - \bar{1} = (x - \bar{1})(x - \bar{2})(x - \bar{3})(x - \bar{4})(x - \bar{5})(x - \bar{6})$.

2. $\bar{2}.\bar{4} = \bar{1}$, $\bar{3}.\bar{5} = \bar{1}$, $\bar{6}.\bar{6} = \bar{1}$.

3. $\bar{3}.\bar{16} = \bar{1}$, $\overline{23}.\overline{45} = \bar{1}$, $\overline{24}.\bar{2} = \bar{1}$, $\overline{32}.\overline{25} = \bar{1}$.

4. (i) $\left.\begin{array}{l} \bar{2}x + \bar{3}y = \bar{1} \\ \bar{3}x + \bar{2}y = \bar{4} \end{array}\right\}$ thus $\left.\begin{array}{l} x + \bar{4}y = \bar{3} \\ x + \bar{4}y = \bar{3} \end{array}\right\}$.

Solutions:

$x = \bar{0}, y = \bar{2}; x = \bar{1}, y = \bar{3}; x = \bar{2}, y = \bar{4}; x = \bar{3}, y = \bar{0}; x = \bar{4}, y = \bar{1}.$

(ii) $\left.\begin{array}{l} x + \bar{3}y = \bar{2} \\ \bar{3}x + \bar{2}y = \bar{2} \end{array}\right\}$ thus $\left.\begin{array}{l} \bar{3}x + \bar{4}y = \bar{1} \\ \bar{3}x + \bar{2}y = \bar{2} \end{array}\right\}$,

Solutions: $\bar{3}.\bar{2}y = \bar{3}.\bar{4}$, giving $y = \bar{2}, x = \bar{1}$.

5. $\left.\begin{array}{l} \overline{15}x + \overline{25}y = \overline{16} \\ \bar{8}x + \overline{21}y = \overline{18} \end{array}\right\}$ thus $\left.\begin{array}{l} \overline{27}x + \overline{14}y = \bar{4} \\ \overline{27}x + \bar{5}y = \overline{22} \end{array}\right\}$,

Solutions: $\bar{7}.\bar{9}y = \bar{7}.\overline{13}$, giving $y = \overline{29}, x = \overline{23}$.

6. $\left.\begin{array}{l} \alpha x + y = \bar{1} \\ x + \beta y = \beta \end{array}\right\}$ thus $\left.\begin{array}{l} \alpha x + y = \bar{1} \\ \alpha x + y = \bar{1} \end{array}\right\}$.

Solutions: $x = \bar{0}, y = \bar{1}; x = \bar{1}, y = \beta; x = \alpha, y = \alpha; x = \beta, y = \bar{0}.$

7. $u, v \in U(R)$, $(uv)^{-1} = v^{-1}u^{-1}$, so $uv \in U(R)$ etc.

8. $F \setminus \{0\}$.

9. For all a, b, $c \in \mathbb{Z}$, $(a + b\sqrt{3}) + (c + d\sqrt{3}) = (a + c) + (b + d)\sqrt{3} \in S$, $-(a + b\sqrt{3}) = (-a) + (-b)\sqrt{3} \in S$, $(a + b\sqrt{3})(c + d\sqrt{3}) = (ac + 3bd) + (ad + bc)\sqrt{3} \in S$. Now $S \subseteq \mathbb{R}$ and so S is a subring of $\mathbb{R}$. S is an integral domain but is not a field.

10. Certainly S is an integral domain. But $0 \neq a + b\sqrt{5} \in S$ implies
$$\frac{1}{a + b\sqrt{5}} = \frac{a - b\sqrt{5}}{a^2 + 5b^2} \in S \text{ and so } S \text{ is a field.}$$

11. S is a field. Prove!

12. Induction: $(a + b)^{p^{n+1}} = ((a + b)^{p^n})^p = (a^{p^n} + b^{p^n})^p$
$= (a^{p^n})^p + (b^{p^n})^p = a^{p^{n+1}} + b^{p^{n+1}}$.
$$\left(\sum_{i=1}^{r} a_i \right)^{p^n} = \sum_{i=1}^{r} a_i^{p^n}.$$

13. Write elements as $\begin{pmatrix} z & w \\ -\overline{w} & \overline{z} \end{pmatrix}$ where $\overline{z}$, $\overline{w}$ are complex conjugates of z, w. Hence prove result.

5.3

1. $a + bi$ is a unit if and only if $a^2 + b^2 = 1$. Thus units are ± 1, $\pm i$.

2. (i) $\dfrac{1 + 13i}{4 - 3i} = (-1 + 2i) + \left(-\dfrac{2}{5} + \dfrac{1}{5}i \right)$.

$q = -1 + 2i$, $r = (1 + 13i) - (4 - 3i)(-1 + 2i) = -1 + 2i$.

(ii) $\dfrac{5 + 15i}{7 + i} = 1 + 2i$.

$q = 1 + 2i$, $r = 0$.

(iii) $\dfrac{5 + 6i}{1 - 3i} = (-1 + 2i) + \left(\dfrac{5}{13} + \dfrac{1}{13}i \right)$.

$q = -1 + 2i$, $r = (5 + 6i) - (2 - 3i)(-1 + 2i) = 1 - i$.

3. For $\alpha, \gamma, \delta \in \mathbb{Z}$,

$$\frac{\alpha + \beta\sqrt{2}}{\gamma + \delta\sqrt{2}} = (m + \varepsilon) + (n + \eta)\sqrt{2} \quad \left(m, n \in \mathbb{Z}, |\varepsilon| \leq \frac{1}{2}, |\eta| \leq \frac{1}{2} \right).$$

Now follow the proof for the Gaussian integers.

4. $w^2 = \overline{w} = -1 - w$, $w\overline{w} = 1$. D is a ring since $(\alpha + \beta w) + (\gamma + \delta w) = (\alpha + \gamma) + (\beta + \delta)w$, $(\alpha + \beta w)(\gamma + \delta w) = (\alpha\gamma - \beta\delta) + (-\beta\delta + \beta\gamma + \alpha\delta)w$ and $D \subseteq \mathbb{C}$

$$\frac{\alpha + \beta w}{\gamma + \delta w} = (m + \varepsilon) + (n + \eta)w \quad \left(m, n \in \mathbb{Z}, |\varepsilon| \leq \frac{1}{2}, |\eta| \leq \frac{1}{2} \right).$$

$r = (\alpha + \beta w) - (\gamma + \delta w)(m + nw)$

$\quad = (\gamma + \delta w)(\varepsilon + \eta w)$.

Now prove $|\varepsilon + \eta w|^2 = \varepsilon^2 + \eta^2 - \varepsilon\eta \leq \frac{3}{4}$ and follow proof for Gaussian integers.

5.4

1. $\begin{pmatrix} x & y \\ 0 & 0 \end{pmatrix} \begin{pmatrix} a & b \\ c & d \end{pmatrix} = \begin{pmatrix} ax + cy & bx + dy \\ 0 & 0 \end{pmatrix}$, etc.

2. The intersection of a collection of left ideals is a left ideal.

3. $x \in R$. $a_1, a_2 \in \bigcup_{i=1}^{\infty} L_i$. For some N, $a_1, a_2 \in L_N$. Thus immediately we have $a_1 + a_2, -a_1, xa_1 \in L_N$. Hence $a_1 + a_2, -a_1, xa_1 \in \bigcup_{i=1}^{\infty} L_i$ which is therefore a left ideal.

4. $0 \in A$ since $x0 = 0$ $(x \in X)$. $y \in R$, $a_1, a_2 \in A$. $x(a_1 + a_2) = xa_1 + xa_2 = 0 + 0 = 0$, $x(-a) = -(xa_1) = 0$, $x(a_1 y) = (xa_1)y = 0y = 0$. Hence A is a right ideal.

5. Since $a = 1a + 0a$, a is in subset. Now prove subset is a left ideal.

6. $M = \{x \in R | f(x) \in L\}$. $y \in R$, $u, v \in M$.
   $f(u + v) = f(u) + f(v) \in L$, $f(-u) = -f(u) \in L$, $f(yu) = f(y)f(u) \in L$.
   Thus we have $u + v, -u, yu \in M$. Hence M is a left ideal.

7. $f(L)$ is certainly a subring of S. Let $a \in f(L)$, $x \in S$. Then $a = f(b)$ $(b \in L)$, $x = f(y)$ $(y \in R)$. $xa = f(y)f(b) = f(yb) \in f(L)$. Thus $f(L)$ is a left ideal of S.

8. $x, y \in S$, $a, b \in I$. $(x + a) + (y + b) = (x + y) + (y + b) \in S + I$, $-(x + a) = (-x) + (-a) \in S + I$, $(x + a)(y + b) = xy + (ay + xb + ab) \in S + I$ (I is an ideal). Thus $S + I$ is a subring. Also $I \cap S$ is a subring of S. If $x \in S$, $a \in I \cap S$, then $xa \in I$ (I left ideal) and $xa \in S$ (S subring). Thus $xa \in I \cap S$ which is a left ideal of S.

9. R, S, T are additive and multiplicative semigroups. Hence result.

10. $(a + ib) + (c + id) = (a + c) + i(b + d) \rightarrow$

$$\begin{pmatrix} a+c & b+d \\ -(b+d) & a+c \end{pmatrix} = \begin{pmatrix} a & b \\ -b & a \end{pmatrix} + \begin{pmatrix} c & d \\ -d & c \end{pmatrix}$$

$(a + ib)(c + id) = (ac - bd) + i(bc + ad) \rightarrow$

$$\begin{pmatrix} ac - bd & ad + bc \\ -ad - bc & ac - bd \end{pmatrix} = \begin{pmatrix} a & b \\ -b & a \end{pmatrix} \begin{pmatrix} c & d \\ -d & c \end{pmatrix}.$$

As mapping is bijective result follows.

5.5

1. M maximal implies R/M is a field and so R/M is an integral domain. Hence M is prime. $\{0\}$ is prime, but not maximal, in $\mathbb{Z}$.

2. $4 + 2i = (1 - 3i)i + (1 + i)$, $1 - 3i = (1 + i)(-1 - 2i)$. Hence it follows that $(4 + 2i)1 + (1 - 3i)(-i) = 1 + i$ (G.C.D.).

3. Certainly $Da_1 + Da_2 + \ldots + Da_n = Dd$ for some d. Then $x_1 a_1 + x_2 a_2 + \ldots + x_n a_n = d$ for some $x_1, x_2, \ldots, x_n \in D$. Prove now d is G.C.D.

5.6

1. Three polynomials are irreducible by Criterion for $p = 5, 7, 13$ respectively.

2. $19 + 24x + 9x^2 + x^3 = 3 + 9y + 6y^2 + y^3$ which is irreducible by Criterion for $p = 3$.

3. $x^3 + 9$ not irreducible in $\mathbb{Q}[x]$ implies that $x^3 + 9$ has a linear factor $x + \alpha$ $(\alpha \in \mathbb{Q})$. Then $\alpha^3 + 9 = 0$ $(\alpha \in \mathbb{Q})$ which is impossible.

6.1

1. $\begin{pmatrix} 1 & 2 & 3 & 4 & 5 & 6 \\ 6 & 2 & 5 & 4 & 1 & 3 \end{pmatrix}, \begin{pmatrix} 1 & 2 & 3 & 4 & 5 & 6 \\ 2 & 4 & 3 & 6 & 5 & 1 \end{pmatrix}.$

2. Direct verification.

3. $(1 \quad 2)(1 \quad 6)(1 \quad 7)(1 \quad 5)(1 \quad 3)$ odd, $(1 \quad 3)(1 \quad 4)(1 \quad 2)$ odd, $(6 \quad 4)(6 \quad 8)(6 \quad 5)(6 \quad 2)(6 \quad 1)(6 \quad 7)$ even.

4. $(1), (1 \quad 2)(3 \quad 4), (1 \quad 3)(2 \quad 4), (1 \quad 4)(2 \quad 3), (1 \quad 2 \quad 3), (1 \quad 3 \quad 2),$
$(1 \quad 4 \quad 2), (1 \quad 2 \quad 4), (2 \quad 4 \quad 3), (2 \quad 3 \quad 4), (1 \quad 4 \quad 3), (1 \quad 3 \quad 4).$

5. Skew-symmetric.

6. $\text{Ker } f = \{a \in G \mid a.x = x \text{ for all } x \in G\} = \bigcap_{x \in X} \text{Stab}_G(x).$

7. $G = a_1\text{Stab}_G(x) \cup a_2\text{Stab}_G(x) \cup \ldots \cup a_n\text{Stab}_G(x)$. Prove orbit of x consists precisely of $a_1.x, a_2.x, \ldots, a_n.x$ and that $a_i.x = a_j.x$ if and only if $a_j^{-1}a_i \in \text{Stab}_G(x)$.

8. $a, b \in G$. $(ab).H_i = (ab)H_i(ab)^{-1} = abH_ib^{-1}a^{-1} = a.(b.H_i)$. $\text{Stab}_G(H_i) = N_G(H_i)$. Hence $n = |G : N_G(H)|$.

9. $e \to (1)$, $\quad a \to (1 \quad 2 \quad 3)$, $\quad b \to (1 \quad 3 \quad 2)$, $\quad c \to (1 \quad 2)$, $\quad d \to (1 \quad 3)$, $\quad f \to (2 \quad 3)$.

6.2

1. $b = a^{-1}$, $a^3 = e$, $b^2 = e$ implies $a = b = e$, $|G| = 1$.

2. From $aba^2 = b$, $b^{-1}ab = a^{-2} = a^2$. Hence $b^{-1}a^2b = a^4 = e$ and so $a^2 = e$. Thus $ab = b$ and $a = e$. $G = \langle b \mid b^2 = e \rangle$.

3. The elements of G are of the form a^n or ba^n $(n = 0, 1, \ldots, 7)$. Thus $|G| = 16$. a^4 has order 2 but $(ba^n)^2 = ba^nba^n = b^2b^{-1}a^nba^n = b^2a^{-n}a^n = b^2 \neq e$ $(n \neq 0)$. Hence a^4 generates only subgroup of order 2.

4. G is the dihedral group D_3.

5. Label vertices consecutively as 1, 2, 3, 4. Group of transformations is $\{(1), (1 \quad 4)(2 \quad 3), (1 \quad 2)(3 \quad 4), (1 \quad 3)(2 \quad 4)\}$.

6. As in text for D_n.

6.3

1. AB is normal. Hence $ABC = (AB)C$ is normal.

2. $G = H \times K$. $a, b \in G$, $a = xy$, $b = zt$ $(x, z \in H; \ y, t \in K)$. $ab = (xz)(yt)$, $ba = (zx)(ty)$. Hence result.

3. Obvious as $H/(H \cap K) \cap K/(H \cap K) = \{H \cap K\}$.

4. No. $G = \langle a, b \mid a + b = b + a, 2a = 2b = 0 \rangle$, $A = \langle a \rangle$, $B = \langle b \rangle$, $C = \langle c \rangle$, where $c = a + b$. Then $G = A + B + C$, $A \cap B = A \cap C = B \cap C = \{0\}$.

6.4

1. $a, b \in M$ implies $ma = 0, nb = 0$ $(m, n \in \mathbb{N})$. $mn(a + b) = n(ma) + m(nb) = 0 + 0 = 0$, $m(-n) = -(ma) = 0$, thus M is a subgroup. Suppose $x \in A, r \in \mathbb{N}$ is such that $r(x + M) = M$. Then $rx + M = M$ and so $rx \in M$. Then for some $s \in \mathbb{N}$, $s(rx) = 0$ and so $(sr)x = 0$ and $x \in M$.

2. $na + nb = n(a + b)$ $(a \in A)$, $-(na) = (-n)a$. Hence nA is a subgroup.

   In solutions 3–5 let C_n be cyclic of order n.

3. $20 = 2^2.5$. $C_2 \times C_2 \times C_5$, $C_4 \times C_5$.

4. $72 = 2^3.3^2$. $C_2 \times C_2 \times C_2 \times C_3 \times C_3$, $C_4 \times C_2 \times C_3 \times C_3$, $C_8 \times C_3 \times C_3$, $C_2 \times C_2 \times C_2 \times C_9$, $C_4 \times C_2 \times C_9$, $C_8 \times C_9$.

5. $2250 = 2.3^2.5^3$. $C_2 \times C_3 \times C_3 \times C_5 \times C_5 \times C_5$, $C_2 \times C_9 \times C_5 \times C_5 \times C_5$, $C_2 \times C_3 \times C_3 \times C_5 \times C_5 \times C_{25}$, $C_2 \times C_9 \times C_5 \times C_{25}$, $C_2 \times C_3 \times C_3 \times C_{125}$, $C_2 \times C_9 \times C_{125}$.

6.5

1. Centre $Z(G) \neq \{e\}$. If $Z(G)$ is not a subgroup of H then $H \subset HZ(G) \subseteq N_G(H)$. If $Z(G) \subseteq H$ then argue by induction on $|G|$. Result is true for $G/Z(G)$. Then $H/Z(G)$ is not the normalizer of $H/Z(G)$ in $G/Z(G)$ from which H is not the normalizer of H in G. Deduction is clear.

2. $|G| = 45 = 3^2.5$. Let m be the number of Sylow 3-subgroups. Then we have $m = 1 + 3r$ and $1 + 3r$ divides $3^2.5$. Hence $r = 0, m = 1$. Let P be the unique Sylow 3-subgroup. Then P has order 9 and so is Abelian. Let n be the number of Sylow 5-subgroups. Then $n = 1 + 3s$ and $1 + 3s$ divides $3^2.5$. Hence $s = 0$, $n = 1$. Let Q be the unique Sylow 5-subgroup. Then Q has order 3 and so is cyclic.
   Since $G = P \times G$ result follows.
   $|G| = 207 = 3^2.23$. Let m, n be the numbers of Sylow 3-subgroups and 23-subgroups respectively. $m = 1 + 3r$, $n = 1 + 23s$ and m and n divide $3^2.23$. Hence $m = 1, n = 1$. G is the direct product of its (Abclian) Sylow subgroups of orders 9 and 23 and so is Abelian.

3. P is a Sylow p-subgroup of G and of PN. Let $|PN| = |P|m, |G| = |P|mn$ where m, n are integers not divisible by p. But PN/N is isomorphic to $P/(P \cap N)$ and so

$$\frac{|PN|}{|N|} = \frac{|P|}{|P \cap N|} \text{ giving } |N| = \frac{|P \cap N||PN|}{|P|} = \frac{|P \cap N||P|m}{|P|} = |P \cap N|m.$$

Thus $P \cap N$ is a Sylow p-subgroup of N. Hence a Sylow p-subgroup of G/N has order $|P : P \cap N| = |PN : N|$. Thus PN/N is a Sylow p-subgroup of G/N.

4. If $p = q$ the group is Abelian and result is true. If $p > q$ let n be the number of Sylow p-subgroups. Then $n = 1 + rp$ and n divides pq. Hence $r = 0$, $n = 1$ and the Sylow p-subgroup is unique.

5. If $1 + 5k$ divides 60 then $1 + 5k$ is 1 or 6.

6. $351 = 3^3.13$. Let n be the number of Sylow 13-subgroups. Then we have $n = 1 + 13k$ and $1 + 13k$ divides $3^3.13$. Hence $1 + 13k = 1$ or 27. If $1 + 13k = 1$ there is a unique Sylow 13-subgroup. If $1 + 13k = 27$ the group contains 27 Sylow 13-subgroups. Then the number of elements in Sylow 13-subgroups is given by $27(13 - 1) + 1 = 325$. Thus there are $351 - 325 = 26$ remaining elements. But a Sylow 3-subgroup has 27 elements including the identity element. Thus the remaining elements must be from a Sylow 3-subgroup which is unique. Hence there is a normal Sylow 13-subgroup or a normal Sylow 3-subgroup.

7. $(13)(24)(12)$ and (12) generate a Sylow 2-subgroup. (Examples 21, 3)

Suggestions for Further Study

1. Beginning Texts

R.A. Dean, *Elements of Abstract Algebra* (Wiley, 1967).

J.B. Fraleigh, *A First Course in Abstract Algebra* (Addison-Wesley, 1989).

J.A. Gallian, *Contemporary Abstract Algebra* (D.C. Heath, 1994).

2. Advanced Texts

P.M. Cohn, *Algebra*, vols. 1–3 (Wiley, 1991) – has an extensive bibliography.

N. Jacobson, *Basic Algebra*, vols. I and II (W.H. Freeman, 1985).

B.L. van der Waerden, *Modern Algebra*, vols. I and II (F. Ungar, 1948) – the classic text and still worth a look.

3. Historical Texts

V.J. Katz, *A History of Mathematics* (HarperCollins, 1993).

B.L. van der Waerden, *A History of Algebra* (Springer-Verlag, 1985).

Index